Lecture Notes in Computer Science   13175

More information about this series at https://link.springer.com/bookseries/558

Éric Renault · Selma Boumerdassi ·
Paul Mühlethaler (Eds.)

Machine Learning for Networking

4th International Conference, MLN 2021
Virtual Event, December 1–3, 2021
Proceedings

Springer

Editors
Éric Renault ⓘ
ESIEE Paris
Noisy-le-Grand, France

Selma Boumerdassi
CNAM
Paris, France

Paul Mühlethaler
Inria
Paris, France

ISSN 0302-9743   ISSN 1611-3349 (electronic)
Lecture Notes in Computer Science
ISBN 978-3-030-98977-4   ISBN 978-3-030-98978-1 (eBook)
https://doi.org/10.1007/978-3-030-98978-1

This Springer imprint is published by the registered company Springer Nature Switzerland AG
The registered company address is: Gewerbestrasse 11, 6330 Cham, Switzerland

Preface

The rapid development of new network infrastructures and services has led to the generation of huge amounts of data, and machine learning now appears to be the best solution to process these data and make the right decisions for network management. The International Conference on Machine Learning for Networking (MLN) aims at providing a top forum for researchers and practitioners to present and discuss new trends in deep and reinforcement learning, pattern recognition and classification for networks, machine learning for network slicing optimization, 5G systems, user behavior prediction, multimedia, IoT, security and protection, optimization and new innovative machine learning methods, performance analysis of machine learning algorithms, experimental evaluations of machine learning, data mining in heterogeneous networks, distributed and decentralized machine learning algorithms, intelligent cloud-support communications, resource allocation, energy-aware communications, software-defined networks, cooperative networks, positioning and navigation systems, wireless communications, wireless sensor networks, and underwater sensor networks.

The fourth edition of the conference took place during December 1–3, 2021. Due to the special health situation all around the world, resulting from the ongoing COVID 19 pandemic, all presentations at MLN 2021 were done remotely.

The call for papers resulted in a total of 30 submissions from all around the world: Canada, China, Egypt, France, Germany, India, Lebanon, South Africa, Sweden, Taiwan, the USA, and Vietnam. All submissions were assigned to at least three members of the Program Committee for review. The Program Committee decided to accept 10 papers.

The paper "DynamicDeepFlow: An Approach for Identifying Changes in Network Traffic Flow Using Unsupervised Clustering" by Sheng Shen, Mariam Kiran, and Bashir Mohammed from the Lawrence Berkeley National Laboratory, USA, was awarded the prize for the best paper.

Four keynotes completed the program: "Distributed and Communication-Efficient ML over and for Wireless" by Mehdi Bennis from the University of Oulu, Finland, "Analysis of Data of Intelligent Transport Systems" by Hacène Fouchal from Université de Reims Champagne-Ardenne, France, "Beyond Shannon: A Theory of Semantic Communication" by Jean-Claude Belfiore from Huawei, France, and "Add Cognitive Capabilities to Apps with Azure Cognitive Services" by Franck Gaillard from Microsoft, France.

We would like to thank all who contributed to the success of this conference. In addition to our roles as General Chairs we chaired the Program Committee together with Selma Boumerdassi, and we thank in particular the members of the Program Committee and the reviewers for carefully reviewing the contributions and selecting a high-quality program. Our special thanks go to the members of the Organizing Committee for their great help.

We hope that all participants enjoyed this successful conference.

December 2021   Paul Mühlethaler
                                                                            Éric Renault

Organization

MLN 2021 was jointly organized by the EVA project of Inria Paris, the Laboratoire d'Informatique Gaspard-Monge (LIGM), ESIEE Paris of Université Gustave Eiffel, and Cnam Paris.

General Chairs

Paul Mühlethaler   Inria, France
Éric Renault   ESIEE Paris, France

Steering Committee

Selma Boumerdassi   Cnam, France
Éric Renault   ESIEE Paris, France

Publicity Chair

Christophe Maudoux   Cnam, France

Organization Committee

Lamia Essalhi   ADDA, France
Nour El Houda Yellas   Cnam, France

Technical Program Committee

Fred Aklamanu	Nokia Bell Labs, France
Anwer Al-Dulaimi	EXFO, Canada
Maxim Bakaev	NSTU, Russia
Jean-François Bercher	ESIEE Paris, France
Selma Boumerdassi	Cnam, France
Nirbhay Chaubey	Ganpat University, India
Alberto Conte	Nokia Bell Labs, France
Aravinthan Gopalasingham	Nokia Bell Labs, France
Viet Hai Ha	Hue University, Vietnam
Rahim Haiahem	Université Badji Mokhtar Annaba, Algeria
Eun-Sung Jung	Hongik University, South Korea
Güneş Karabulut-Kurt	Polytechnique Montréal, Canada

Cherkaoui Leghris	Hassan II University, Morocco
Olaf Maennel	Tallinn University of Technology, Estonia
Ruben Milocco	Universidad Nacional del Comahue, Argentina
Paul Mühlethaler	Inria, France
Gianfranco Nencioni	University of Stavanger, Norway
Sabine Randriamasy	Nokia Bell Labs, France
Éric Renault	ESIEE Paris, France
Stavros Shiaeles	University of Portsmouth, UK
Abdallah Sobehy	Amazon, UK
Van Long Tran	Phu Xuan University, Vietnam
Haibo Wu	Chinese Academy of Sciences, China
Mohamed Faten Zhani	ETS Montreal, Canada

Sponsoring Institutions

Cnam, Paris, France
ESIEE, Paris, France
Inria, Paris, France
Université Gustave Eiffel, France

Contents

Evaluation of Machine Learning Methods for Image Classification: A Case Study of Facility Surface Damage

Ching-Lung Fan(✉) (iD)

Department of Civil Engineering, The Republic of China Military Academy, No. 1, Weiwu Rd., Fengshan, Kaohsiung 830, Taiwan
p93228001@ntu.edu.tw

Abstract. Common reinforced concrete (RC) damage includes exposed rebars, spalling, and efflorescence, which not only affect the aesthetics of facilities but also cause structural degradation over time, setting the stage for further severe RC degradation that would reduce the strength and durability of the structure. Damage to RC facilities occurs because of their natural deterioration. Machine learning can be employed to effectively identify various damage areas, and the findings can serve as a reference to management units in the task of ensuring the structural safety of facilities. In this study, a damage image was used to evaluate image classification capabilities achievable through maximum likelihood and random forest supervised machine learning methods. With these methods, accuracies of 98.6% and 96% were achieved for RC damage classification, respectively. The results of this study demonstrate that the use of machine learning can yield favorable results for damage image classification.

Keywords: Machine learning · Maximum likelihood · Random forest · Image classification · Damage

1 Introduction

Computer vision technology has been recognized as a key component that can help improve detection and monitoring in the field of civil engineering. It has the potential to be used in combination with visual methods and digital cameras for the rapid and automatic inspection and monitoring of civil infrastructure conditions [1]. Valero et al. [2] explained that red, green, and blue (RGB) data obtained with these devices can be processed using different mathematical operations to encode and quantify changes in materials and that these operations can be used for detection and classification purposes. Compared with conventional manual methods, the use of computer vision and image processing technologies can provide accurate results and help overcome the problems of subjective judgment and deviation in the visual inspection of surfaces. These technologies differ in their complexity and approach to image processing. In particular, machine learning is a popular technology that enables computers to learn about certain tasks from a large amount of data and to assess the performance of such tasks; as the machine becomes

© Springer Nature Switzerland AG 2022
É. Renault et al. (Eds.): MLN 2021, LNCS 13175, pp. 1–10, 2022.
https://doi.org/10.1007/978-3-030-98978-1_1

more experienced, the technology improves. Methods based on machine learning have been used to detect various types of RC damage, including spalling [3–5], exposed rebars [6, 7], and cracking [8–10].

Early damage detection can facilitate the adoption of measures to prevent damage and possible structural failure. A combination of visual inspection and measurement tools can facilitate an assessment of the inadequacies of surface conditions [11]. To replace inefficient manual inspection methods, researchers have developed automated inspection methods for effectively detecting RC damage. Machine learning is an emerging approach that can help rapidly and accurately identify damage in images, and it has gained widespread attention. In studies on damage identification in concrete structures, machine learning methods have been used to analyze existing inspection results, extract damage characteristics, and reflect these characteristics in various damage classifications. Amezquita-Sanchez and Adeli [12] remarked that machine learning is suitable for damage detection. Examples of such use are structural health monitoring of structures such as concrete dams [13], structural health monitoring and damage mode of buildings [14], and seismic assessment of school building structures [15].

The objective of this study was twofold: to use low-cost and convenient images for testing machine learning without implementing artificial image preprocessing to determine the classification results for damage images and to provide a reference for improving image detection methods for use in assessing the condition of automated RC facilities.

2 Research Methodology

Machine learning methods are highly advanced tools that use the extracted features of images to perform specific tasks, such as classification, regression, and clustering [16]. Classification is a machine learning task aimed at determining the relationships between objects and labels through exposure to a training set, which is a collection of objects with known labels [17]. This approach can be applied to classify objects with unknown labels in future applications. Machine learning entails the process of a machine learning a set of rules from instances (training set samples) and it subsequently using a classifier to apply these rules to new instances [18]. In general, machine learning can be divided into supervised and unsupervised learning. Supervised learning involves learning a model (classification or regression model) from training data with labels and predicting unknown samples according to the model. The input of the model is the sample features, and the output of the function is the label corresponding to the sample. The input samples in unsupervised learning do not need to be labeled, but the machine learns the features from the samples to make predictions. It can automatically subject the input data to classification or clustering without being given any labeled training examples in advance. Due to the lack of category labels, unsupervised learning requires the determination of the optimal feature vector for data classification [19].

The tasks of supervised learning include the use of classification and regression models. Supervised learning involves a classification model if the output consists of discrete data, whereas a regression model is used if the output contains continuous data. Both models share the same purposes of predicting, classifying, identifying, and detecting unknown data and are applied for feature classification, image recognition, and damage detection. Classification (i.e., classifier) model algorithms include those for Naïve Bayes, maximum likelihood (ML), and logistic regression; regression models use algorithms such as linear regression, k-nearest neighbor (k-NN), and AdaBoost. In addition, support vector machine (SVM), random forest (RF), and classification and regression tree (CART) models can all perform classification and regression; accordingly, a classification or regression model is selected based on the data type collected (continuous or discrete values). Subsequently, users can evaluate the performance of the model based on the results of the confusion matrix. The tasks of unsupervised learning include clustering and association rules, with common algorithms being K-means, Apriori, self-organizing map, and principal components analysis. The difference between clustering and classification is that clustering is a set of unsupervised learning methods that do not require prelabeled data sets to train the model, whereas classification involves a set of supervised learning methods that require prelabeled data sets to train the model [20].

Machine learning can capture parameter patterns of complex connections hidden in a large amount of data, and therefore, it is suitable for constructing classification models for concrete degradation [21]. Two machine learning algorithms were used in this study, and they are described in this section.

2.1 Maximum Likelihood

Maximum likelihood (ML) is one of the most commonly used methods in image classification, and it involves the assumption that the sample data have a normal distribution. A supervised classification method derived from Bayes' theorem, its principle is that pixels that have the greatest likelihood of belonging to a category are classified into the category. The posterior distribution $P(i \mid \omega)$ is the probability that a pixel with feature vector ω belongs to category i, and it is expressed as

$$p(i|\omega) = \frac{p(\omega|i)p(i)}{P(\omega)} \qquad (1)$$

where $P(\omega|i)$ is the likelihood function, $P(i)$ is a priori information, specifically the probability that category i appears in the research area, and $P(\omega)$ is the probability of observing ω and is usually regarded as a normalization constant. ML classification involves the assumption that the statistical data of all categories in each band are normally distributed, and the calculation of the probability that a given pixel belongs to a specific category. Each pixel is assigned to the category for which the pixel has the highest probability of belonging (i.e., the ML). If the highest probability is less than the specified threshold, the pixel remains unclassified or unknown. Therefore, ML classification of an image can be performed by calculating the probability density function, given by Eq. (2) below, of each pixel in the image [22].

$$P(x|i) = (2\pi)^{-n/2} \left| \sum_i \right|^{-1/2} \exp\left[-\frac{1}{2}(x - mi)^T \sum_i^{-1} (x - mi)\right] \qquad (2)$$

2.2 Random Forest

Because a single classifier sometimes cannot effectively process classification problems, combining several learning methods is necessary to obtain more satisfactory results compared with the use of a single method, and to improve the robustness and accuracy of the model. This group of multiple models is called an ensemble, and the technique is called ensemble learning [23]. Random forest (RF), proposed by Breiman [24], involves the use of a combination of bagging and random subspace methods to construct multiple decision trees for classification or regression, after which voting or averaging is performed using the predicted values of each decision tree to obtain the final classification result.

RF is based on the concept of classification and regression trees, where a robust model is constructed by combining multiple weak learners. When the ensemble model constructs a decision tree, samples are retrieved from the training set and then placed back in it. From several variables randomly selected from a random subset, the optimal variable is selected and used as the branch attribute. Nonlinear decision boundaries are created through ensemble learning. This is a technique for training multiple decision trees on a random subset of the training data. The multiple decision trees created and trained represent a forest system, and the final classification is a statistical model of tree collection and prediction [25]. Athanasiou et al. [8] remarked that ensemble learning tends to provide more favorable performance for classification models by eliminating the noise (caused by a specific data set) in each model. Therefore, in the RF algorithm, the large number of trees do not cause overfitting, and choosing the correct type of random variable enables accurate classification [26].

2.3 Evaluation Index of Classification Model

The performance of the classification model for supervised machine learning (referring to the generalization capability of image damage recognition) had to be measured by a certain index, and the parameters of the machine learning algorithms or the variables of the classification target could be adjusted through the indicators to gradually optimize the model. A confusion matrix is the most common method for assessing classification models. It is presented in a specific tabular format to understand whether the model confuses different categories through a visualized approach and is used to evaluate the classification performance of supervised machine learning. The evaluation method involves comparing the data predicted by the classification model with the actual data to measure the classification results of the model.

Use of a confusion matrix assumes that a classification model has two categories: X and Y. If the prediction result of a model is consistent with the actual classification of the data, the result is true (i.e., a correct prediction). There are two types of correct predictions; the first involves both an actual classification and a prediction of X, namely a true positive (TP); the second correct prediction consists of both an actual classification and a prediction of Y, namely a true negative (TN). If the actual classification and prediction are inconsistent, the prediction is false, a misjudgment. There are two types of misjudgments, the first of which involves an actual classification of X and a prediction of Y, namely a false negative (FN); the second prediction consists of an actual classification of Y and a prediction of X, and this is known as a false positive (FP) (Table 1).

– Accuracy

The accuracy of classification is the main consideration in the detection system, and the recognition feature performance of surface defects is the basis of the system [27]. Kotsiantis [18] indicated that the evaluation of classifiers is usually based on prediction accuracy (the percentage of correct predictions divided by the total number of predictions) and compared supervised learning algorithms by using the accuracy of the trained classifiers on a specific data set. Accuracy refers to the ratios of all prediction categories that actually belong to a particular category, as shown in (3). However, accuracy is not the most reliable indicator for the actual evaluation of the classifier because if the data set is unbalanced, this produces misleading results when the number of samples in different categories varies greatly.

$$\text{Accuracy} = \frac{TP + TN}{TP + FN + FP + TN} \tag{3}$$

– Precision and Recall

Although accuracy can be a measure of the quality of a classification model, it is incapable of meeting all evaluation requirements. It can be used as an evaluation index by calculating precision and recall. Precision reveals how many of the predicted categories actually belong to a particular category (4), and a higher precision leads to a lower rate of misjudgment. Because precision only involves making positive predictions, the classifier would ignore all but one positive instance, and thus, precision has to be used together with another evaluation indicator, namely recall. Recall is the actual ratio of data in a certain category to be predicted to belong to that category (5), meaning the ratio of positive instances that are correctly predicted.

$$\text{Precision} = \frac{TP}{TP + FP} \tag{4}$$

$$\text{Recall} = \frac{TP}{TP + FN} \tag{5}$$

Table 1. Confusion matrix of classification model

Actual	Predicated	
	X	Y
X	True Positive (TP)	False Negative (FN)
Y	False Positive (FP)	True Negative (TN)

– F1

Precision (P) and recall (R) can only reflect a certain aspect of a problem; thus, evaluating the classification model through P or R alone may result in inaccuracy. Therefore, F1 is used as the harmonic mean of P and R, as shown in Eq. (6) to achieve a balance between P and R (the two are usually inversely proportional) and to enable both to achieve their maximum value. When using a harmonic mean, more weight is assigned to lower values; therefore, if P and R are both high, the classifier's F1 will receive a higher score. F1 can explain the weighted average of P and R, and it reaches its optimal and worst values at 1 and 0, respectively [28].

$$F1 = \frac{2PR}{P+R} \tag{6}$$

3 Analytical Results

Noncontact technologies can facilitate the nondestructive evaluation of fabrics and yield accurate and precise geometric- and color-related data, which can be processed to acquire meaningful information [2]. Because the objects detected in digital camera images were identified using noncontact methods, some of the limitations faced during the use of contact sensors for inspection could be easily avoided. Furthermore, the image resolution of recent digital cameras exceeds 10 million pixels, and hence, detailed images of concrete surfaces can be acquired with these cameras [29]. Concrete surfaces can be captured in one shot with a commercial camera, which offers an inexpensive way to obtain a wide range of images for inspection purposes [30]. The size of each color image recorded by a digital camera in this study was 3264 pixels × 2448 pixels, a resolution that was sufficient to identify the photographed objects. The images depicted three types of RC damage: exposed rebars, spalling, and efflorescence. The images comprised an orthoimage (the optical axis of the camera was perpendicular to the RC damage surface) with uniform illuminance.

In this study, the ability of machine learning to recognize RC damage can be understood, avoiding the probability of detection errors. According to Table 2, the accuracy of ML and RF f was 98.6% and 96.0%, respectively. The number of test samples used for detecting RC damage was 500. Table 3 shows the predicted and actual situations of exposed rebars, spalling, and efflorescence. The two classifiers yielded almost 100% accurate predictions for exposed rebars, and they provided optimal classification results despite the sample size of this type of damage being relatively small in the data. RC damage classification with ML and RF showed accuracies of 98.6% and 96%, respectively. Figure 1 shows the results of a visual approach that was used to understand the differences between ML and RF for the three RC damage classifications. A careful analysis can be conducted against the original images. In particular, Fig. 1(c) clearly indicates that RF predicted spalling as rebar (red oval), and Table 3 shows the total number of spalling cases to be 171, with seven cases being misjudged as exposed rebar. Spalling and efflorescence were two damage types that were prone to misjudgment, with both classifiers predicting efflorescence as spalling.

The predicted precision and recall values of the classifiers served as the ranking, and a P–R graph was plotted with precision as the vertical axis and recall as the horizontal axis. When positive was relatively less or more attention was given to false positive instead of false negative, the P–R curve effectively demonstrated that the classifiers had room for improvement (the curve could be closer to the upper left corner). In addition, the graph could visually display the precision and recall of the classifiers in the overall sample. If

Table 2. Evaluation images of RC damage recognition

Machine learning	Accuracy	Precision	Recall	F1
ML	0.986	0.961	0.992	0.976
RF	0.960	0.839	0.943	0.888

(a) Original image

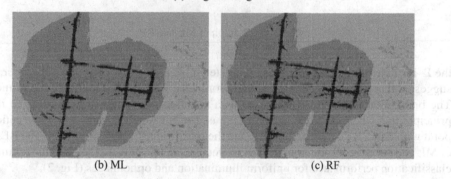

(b) ML   (c) RF

Fig. 1. RC damage classification result for images

Table 3. Confusion matrix of ML and RF for damage image

Actual		Predicated		
		Exposed rebar	Spallig	Efflorescence
ML	Exposed rebar	11	0	0
	Spallig	1	170	0
	Efflorescence	0	6	312
RF	Exposed rebar	10	1	0
	Spallig	7	163	1
	Efflorescence	0	11	307

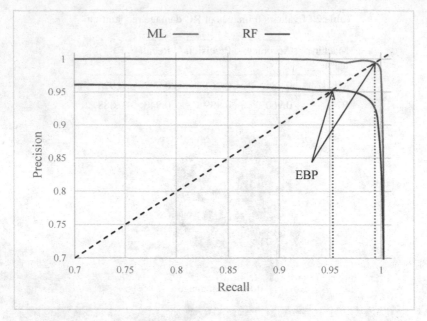

Fig. 2. P–R graph of ML and RF

the P–R curve of one classifier was completely covered by that of another classifier, it suggested that the latter's classification performance was superior to that of the former. The break-even point (BEP) was used as a means of measurement according to the principle that when the values of precision and recall were the same, the intersection point with the P–R curve was the BEP, and the recall value of BEP was taken. The BEPs of ML and RF were 99.43% and 95.3%, respectively, indicating that ML had the optimal classification performance for uniform illumination and orthoimages (Fig. 2).

4   Conclusion

RC damage is inevitable in the natural process of structural wear and may cause severe damage to the overall structure. Despite a lack of theories supporting an optimal method for RC damage detection, classification, and quantification, current machine learning methods have made considerable progress in image classification. In terms of speed, effectiveness, and economy, machine learning has high potential for use in the classification of RC damage images. It can significantly reduce the time and cost of RC facility inspection. The results of facility testing are very important information and can be used as a basis for facility maintenance and limited resource allocation. In the future, machine learning can help overcome the subjectivity of human visual judgment, which causes inconsistencies in the application of standards. This would lead to judgments based on automated image classification having a high level of credibility.

The two main frameworks for automated RC damage inspection are image data acquisition by remote sensing and data processing through computer vision technology. Although the digital cameras, drones, and robotic systems for capturing images are relatively mature and convenient, image quality is affected by the difficult-to-control shooting environment, which can interfere with the data processing and recognition capabilities of machine learning to varying degrees. The findings can serve as a reference for optimizing machine learning parameters, which may facilitate overcoming the limitations in current machine vision computing and thereby improving automated machine learning and recognition capabilities in the future.

References

1. Spencer, B.F., Hoskere, V., Narazaki, Y.: Advances in computer vision-based civil infrastructure inspection and monitoring. Engineering 5(2), 199–222 (2019)
2. Valero, E., Forster, A., Bosché, F., Hyslop, E., Wilson, L., Turmel, A.: Automated defect detection and classification in ashlar masonry walls using machine learning. Autom. Construct. 106, 102846 (2019)
3. Dawood, T., Zhu, Z., Zayed, T.: Machine vision-based model for spalling detection and quantification in subway networks. Autom. Constr. 81, 149–160 (2017)
4. Guldur Erkal, B., Hajjar, J.F.: Laser-based surface damage detection and quantification using predicted surface properties. Autom. Constr. 83, 285–302 (2017)
5. Mizoguchi, T., et al.: Quantitative scaling evaluation of concrete structures based on terrestrial laser scanning. Autom. Constr. 35, 263–274 (2013)
6. Hüthwohl, P., Lu, R., Brilakis, I.: Multi-classifier for reinforced concrete bridge defects. Autom. Construct. 105, 102824 (2019)
7. Rubio, J.J., et al.: Multi-class structural damage segmentation using fully convolutional networks. Comput. Indust. 112, 103121 (2019)
8. Athanasiou, A., Ebrahimkhanlou, A., Zaborac, J., Hrynyk, T., Salamone, S.: A machine learning approach based on multifractal features for crack assessment of reinforced concrete shells. Comput. Aided Civil Infrastruct. Eng. 35(6), 565–578 (2020)
9. Okazaki, Y., Okazaki, S., Asamoto, S., Chun, P.: Applicability of machine learning to a crack model in concrete bridges. Comput. Aided Civil Infrastruct. Eng. 35(8), 775–792 (2020)
10. Li, G., Zhao, X., Du, K., Ru, F., Zhang, Y.: Recognition and evaluation of bridge cracks with modified active contour model and greedy search-based support vector machine. Autom. Constr. 78, 51–61 (2017)
11. Dhital, D., Lee, J.R.: A fully non-contact ultrasonic propagation imaging system for closed surface crack evaluation. Exp. Mech. 52, 1111–1122 (2012)
12. Amezquita-Sanchez, J.P., Adeli, H.: Synchrosqueezed wavelet transform-fractality model for locating, detecting, and quantifying damage in smart highrise building structures. Smart Mater. Struct. 24(6), 065034 (2015)
13. Kang, F., Li, J., Dai, J.: Prediction of long-term temperature effect in structural health monitoring of concrete dams using support vector machines with Jaya optimizer and salp swarm algorithms. Adv. Eng. Softw. 131, 60–76 (2019)
14. Chang, C.M., Lin, T.K., Chang, C.W.: Applications of neural network models for structural health monitoring based on derived modal properties. Measurement 129, 457–470 (2018)
15. Chi, N.W., Wang, J.P., Liao, J.H., Cheng, W.C., Chen, C.H.: Machine learning-based seismic capability evaluation for school buildings. Autom. Construct. 118, 103274 (2020)
16. Zhang, C., Chang, C.C., Jamshidi, M.: Concrete bridge surface damage detection using a single-stage detector. Comput. Aided Civil Infrastruct. Eng. 35, 389–409 (2020)
17. Meijer, D., Scholten, L., Clemens, F., Knobbe, A.: A defect classification methodology for sewer image sets with convolutional neural networks. Autom. Constr. 104, 281–298 (2019)

18. Kotsiantis, S.B.: Supervised machine learning: a review of classification techniques. Informatica **31**, 249–268 (2007)
19. Yuwono, M., et al.: Unsupervised feature selection using swarm intelligence and consensus clustering for automatic fault detection and diagnosis in heating ventilation and air conditioning systems. Appl. Soft Comput. **34**, 402–425 (2015)
20. Dong, C.Z., Catbas, F.N.: A review of computer vision–based structural health monitoring at local and global levels. Struct. Health Monit. **20**(2), 692–743 (2020)
21. Taffese, W.Z., Sistonen, E.: Machine learning for durability and service-life assessment of reinforced concrete structures: recent advances and future directions. Autom. Constr. **77**, 1–14 (2017)
22. Richards, J.A.: Remote Sensing Digital Image Analysis. Springer, Heidelberg (2013). https://doi.org/10.1007/978-3-642-30062-2
23. Rokach, L.: Ensemble-based classifiers. Artif. Intell. Rev. **33**, 1–39 (2010)
24. Breiman, L.: Random forests. Mach. Learn. **45**, 5–32 (2001)
25. Castagno, J., Atkins, E.: Roof shape classification from LiDAR and satellite image data fusion using supervised learning. Sensors **18**(11), 3960 (2018)
26. Yaseen, Z.M., Ali, Z.H., Salih, S.Q., Al-Ansari, N.: Prediction of risk delay in construction projects using a hybrid artificial intelligence model. Sustainability **12**, 1514 (2020)
27. Zhou, S., Chen, Y., Zhang, D., Xie, J., Zhou, Y.: Classification of surface defects on steel sheet using convolutional neural networks. Mater. Technol. **51**, 123–131 (2017)
28. Cho, G.S., Gantulga, N., Choi, Y.W.: A comparative study on multi-class SVM & kernel function for land cover classification in a KOMPSAT-2 image. KSCE J. Civ. Eng. **21**(5), 1894–1904 (2017)
29. Mohan, A., Poobal, S.: Crack detection using image processing: a critical review and analysis. Alex. Eng. J. **57**, 787–798 (2018)
30. Rodríguez-Martín, M., Lagüela, S., González-Aguilera, D., Martínez, J.: Thermographic test for the geometric characterization of cracks in welding using IR image rectification. Autom. Constr. **61**, 58–65 (2016)

One-Dimensional Convolutional Neural Network for Detection and Mitigation of DDoS Attacks in SDN

Abdullah Alshra'a$^{(\boxtimes)}$ (iD) and Seitz Jochen (iD)

Communication Networks Group, Department of Electrical Engineering and
Information Technology, Technische Universität Ilmenau, Ilmenau, Germany
{abdullah.alshraa,Jochen.seitz}@tu-ilmenau.de

Abstract. In Software-Defined Networking (SDN), the controller plane
is an essential component in managing network traffic because of its
global knowledge of the network and its management applications. However, an attacker might attempt to direct malicious traffic towards
the controller, paralyzing the entire network. In this work, a One-
Dimensional Convolutional Neural Network (1D-CNN) is used to protect the controller evaluating entropy information. Therefore, the CICD-
DoS2019 dataset is used to investigate the proposed approach to train
and evaluate the performance of the model and then examine the effectiveness of the proposal in the SDN environment. The experimental
results manifest that the proposed approach achieves very high enhancements in terms of accuracy, precision, recall, F1 score, and Receiver Operating Characteristic (ROC) for the detection of Distributed Denial of
Service (DDoS) attacks compared to one of the benchmarking state of
the art approaches.

Keywords: Software-Defined Networking (SDN) · Intrusion Detection
System (IDS) · Convolutional Neural Network (CNN) · Distributed
Denial of Service (DDoS) · Deep Learning Algorithm (DLA)

1 Introduction

In traditional networks, the forwarding device runs autonomously without con-
cern about how the other devices treat the traffic. Hence, each device has its
routing algorithm to forward the traffic. Once the network administrators decide
to change the network policy or any aspects in the routing algorithm, the configuration process is achieved individually [15]. Therefore, Software-Defined Networking (SDN) solves the programmable network problem via separating the
controller plane from the data plane. As a consequence, the control plane (SDN
controllers) interacts with data plane (switches that are SDN-enabled) using the
OpenFlow protocol through the connection channel or the southbound interface
between the controller and SDN switches [7,11].

© Springer Nature Switzerland AG 2022
É. Renault et al. (Eds.): MLN 2021, LNCS 13175, pp. 11–28, 2022.
https://doi.org/10.1007/978-3-030-98978-1_2

SDN introduces many benefits, such as minimizing operating and maintaining costs, fine-grained traffic control, or fast service deployment. The controller has applications configured according to the administrator's policies. These applications aim to manage the network traffic and make a decision on behalf of the data plane in case a new flow reaches the switch with new features. Each switch has a flow table with the installed instructions. However, if the switch receives a packet that does not match any entry in the flow table, the switch uses a *Miss Table* entry which forwards the received packet as *Packet_In* message to the controller. After that, the controller extracts the packet header and obtains the essential features for the routing algorithm to respond with a *Packet_Out* message containing the proper rule to be installed in the requesting switch's flow table [19].

Although the communication process between the controller and the switches is effective and enhances the flexibility and programmability, it provides new vulnerability and security challenges. The adversary exploits this process and crafts a massive amount of packets with different features (e.g., IP and MAC addresses). This way, it hardly matches any installed flow entry and hence the *Miss Table* will forward the unmatched packets to the controller. This type of attack targets to consume control channel bandwidth and limited resources in both controller and switches, and is a type of Denial of Service (DoS) attack known as the saturation or injection attack [14]. Nevertheless, legacy approaches for detecting Distributed Denial of Services (DDoS) attacks are typically human intervention, including access control and relatively high labor cost, or Intrusion Detection Systems (IDSs) with a high false alarm rate. Thus, this is not suitable for handling the SDN injection attack and cannot ideally differentiate benign traffic from malicious traffic. In addition, with scaling up the internet or any other network type, the possibility to launch a malicious attack is being increased to destroy the performance of the network components such as end devices, servers, mobile devices, electronic systems, and even the transmitted data. Therefore, it is essential to develop IDSs which enable the network administrators to protect their networks. That means IDSs have to be highly efficient to classify the network traffic flows as benign or malicious by extracting and analyzing the traffic features based on predefined rules [15]. However, the attackers always enhance their methods to align with the IDSs rules as much as possible and generate traffic with the same benign traffic characteristics. As a result, the new malicious traffic could not be detected.

Typically, Machine Learning (ML) and Deep Learning (DL) algorithms have been used in many IDSs patterns because of their effectiveness in the classification of the network activities (e.g., Support Vector Machine (SVM), Logistic Regression (LR), Naïve Bayes Machine (NBM), etc.). Machine Learning is a shallow learning, where the algorithms have predefined characteristics for benign or malicious traffic. Moreover, Machine Learning mostly depends on the linear conduct of the data to have a perfect basis for classification. It is vital to select the features that do not have the behavior of non-linearity because this will lead to low accuracy. As well, some features could be remarkable characteristics for a specific attack type and irrelevant to other attack types [16]. Therefore, Deep Learning Algorithms (DLAs) come as a perfect solution with high accuracy and performance for traffic classification. The DLAs can extract the features and classify the traffic without previous knowledge or need to identify the traffic behavior. DLAs use the concept of neural networks in the human brain. They recognize problems utilizing running queries through variant layers according to related mathematical equations to find appropriate identifiers to classify traffic as benign or malicious [10,24].

In this paper, the One-Dimensional Convolutional Neural Network (1D-CNN) is applied as the central part of our IDS that combines the entropy calculation and a Convolutional Neural Network (CNN) algorithm to detect an injection attack in an SDN environment. The remainder of the paper is organized as follows. Section 2 presents some related work in line with the paper subject. In Sect. 3, the 1D-CNN is introduced. Section 4 explains our approach and Sect. 5 discusses its evaluation with the CICDDoS2019 dataset. Finally, Sect. 6 summarizes the paper and shows the future direction of the work.

2   Related Work

Over the last decade, many researchers have been proposing SDN as a perfect environment to achieve network security by considering its advantages [8,9]. Moreover, SDN security has been a hot research topic and still needs innovative suggestions to treat the new vulnerabilities that appeared with the SDN structure. This section presents a brief overview of recent and popular work that suggested dealing with DDoS attacks in SDN environments.

The researchers in [22] introduce the CICDDoS2019 dataset and applied four machine learning algorithms (Random Forest (RF), Multinomial Logistic Regression, Naïve Bayes, ID3) to classify the dataset traffic flows. The scored result did not pass 80% in the best case. In [29], Ye proposes using SVM to protect the SDN controller against DDoS attacks. The scenario of the work was emulated by Mininet and divided into two phases, training and test, where the controller extracts six features from the *Packet_In* message to detect three types of DDoS attacks.

Recently, DLAs have been used and developed rapidly; thus, many scientists start implementing DLAs for DDoS attack detection. In [24], the authors suggested a Gated Recurrent Units Recurrent Neural Network (GRU-RNN) as IDS to work in the SDN environment. The NSL-KDD dataset[1] is used with selected six features in line with the SDN *Packet_In* message. The suggestion obtains an improvement compared to some algorithms with a detection rate of 89%. Hu et al. [17] used an adaptive synthetic sampling (ADASYN) algorithm in order to enhance CNN performance. ADASYN adds new samples for minor classes to handle unbalanced data. Thereby, the solution stops the model to be biased towards frequent samples. Furthermore, the split convolution module SPC-CNN method is applied to defeat the problem of inter-channel information

---

[1] https://www.unb.ca/cic/datasets/nsl.html.

redundancy and improve the diversity of the features. The enhanced CNN model achieved 83.83% accuracy on the NSL-KDD testing data. The enhancement is 4.35% higher than the legacy CNN. Xiao et al. [26] implemented a CNN model with two convolutional layers, max pooling, and one fully connected layer. Also, the work used the Principal Component Analysis (PCA) and the autoencoder reduction techniques to decrease the feature of the input data. PCA reduced the features to 100 and 121, while the autoencoder provides different feature numbers of 100, 121, 81, and 64. The work used the KDDcup99 dataset, and the results proved that the autoencoders are more effective in feature reduction than the PCA technique.

The research work in [18] proposes two detection solutions that stop DDoS attacks in SDN, where snort collects information about the network traffic and SVM or Deep Neural Network algorithms (DNN) are applied for traffic classification after training with the KDDCUP'99 dataset. The work obtains 74.30% accuracy for SVM and 92.30% for DNN. In [21], Prasath proposed an agent program to prevent an external attack against SDN switches. A metaheuristic bayesian network classification algorithm is used to classify the traffic into benign or malicious. The research work scored 82.99% overall accuracy, 77% precision, 74% and 75% for recall and F1 score respectively. In [13], the author presents a method to detect DDoS attacks in an SDN through machine learning and a statistical method. The work comprises three sections: collector, entropy-based, and classification sections based on different machine learning algorithms. The work achieves good accuracy in detecting DDoS attacks in SDN, where several datasets are applied: the UNB-ISCX, CTU-13, and ISOT datasets.

Other research approaches use entropy as the main metric to classify packets as DDoS attacks on an SDN. Swami et al. [23] consider entropy as the main value to deal with *Packet_In* as malicious traffic. Furthermore, the window size to calculate the entropy equals a certain number of packets and their respective destination IP addresses. Xu et al. [27] define the sensitive fields of *Packet_In* messages via a statistical analysis method, which helps the SDN controller obtain the entropy value and determine if there is a presence of a DDoS attack based on violating the dynamic threshold. Moreover, the work adopts ID3 concepts to calculate the information gain of the port source of the attack and limits the rate of these ports. However, entropy-based solutions cannot provide a guaranteed recognition of a DDoS attack, but they are good for assessing the network traffic behavior. Additionally, it should be pointed out that entropy does not consider various factors when calculating the probability distribution of a certain traffic feature. In [12], the authors introduce an enhanced controller application that uses a dynamic entropy threshold to protect the network resources such as hosts, switches, and the SDN controller against DDoS attacks. The work deals with the possibility to craft both IP and MAC addresses. However, the research depends on changing some structure of SDN switches and transferring some of the controller abilities to let the SDN switches monitor the traffic and notify the SDN controller as long as there is traffic violating the threshold.

3   One Dimensional Convolutional Neural Network (1D-CNN)

Convolutional Neural Network (CNN) is a popular DLA for image processing applications. It proves its success compared to the other algorithms in different areas that rely on sequential data such as audio, time series, and natural language processing. The convolution term is a mathematical equation working to combine two functions to produce a third function that aims to merge two sets of information [13]. In general, there are in three different CNN types: one-dimensional convolution, mainly used for sequential input, such as text or audio. Secondly, two-dimensional convolution is used where the input is an image. Thirdly, three-dimensional convolution is applied in three-dimensional medical imaging or detecting.

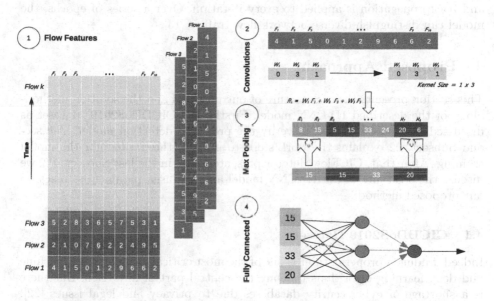

Fig. 1. One-Dimensional Convolutional Neural Network technique

In our work, one-dimensional convolution is applied to analyze time-series information describing a traffic flow received at a specific port, device, or any application. As it is shown in Fig. 1, each traffic flow contains many packets, which creates the features of the flow. Every feature has a value considered as the input value for the 1D-CNN algorithm. 1D-CNN takes the values of the feature and scans them with a filter called kernel size $(X \times Y)$. Then, the filter slides over the feature values, the dot product is calculated between the filter and the parts of the features concerning the size of the filter $(X \times Y)$. After that, the pooling layer uses another filter to select the maximal value or average value for the filter size to decrease the size of the input features [28]. Consequently, the pooling layer decreases the computational power required to process the data through dimensional reduction. Also, the most influential features are defined, which maintains the process of effectively training the model by a particular equation such as a *Rectified Linear Unit* (ReLU) [15]. The outcome of the previous layer is either used for other convolution and pooling layers or distributed to all neural nodes in a fully connected layer, where each value has a certain weight. Thereby, these values are combined to use the result of the equation. Adding a fully connected layer is a perfect method for treating non-linear values and learning a possibly non-linear function in that space. 1D-CNN would flatten the result of the last pooling layer into a column vector to feed all neural nodes in the fully connected layer. In the end, the output layer classifies the flow as benign or malicious according to the *Sigmoid* classification technique or could classify the flow into a particular type of attack by using *Softmax* classification technique [15]. Moreover, the output is fed into a feed-forward neural network, and backpropagation is applied to every iteration. Over a series of epochs, the model can distinguish diverse network flow traffics [13].

4 Proposed Approach

This section presents the architecture of our proposal and the solution methodology of the described 1D-CNN model. Firstly, the CICDDoS2019[2] dataset as discussed in Sect. 4.1 is used to verify the proposed detection model. The second Subsect. 4.2 explains the work steps to prepare the dataset for the model training. After that, CICFlowMeter is presented and, in the last Subsect. 4.3, we discuss the structure of the 1D-CNN model and the used parameters to set up our proposed method.

4.1 CICDDoS2019 Dataset

Indeed, finding a proper dataset is one of the most critical challenges for machine and deep learning algorithms that are the central part of any IDSs. But there is a shortage of cybersecurity databases due to privacy and legal issues [22]. For example, network traffic has critical information that is not reasonable to make public (e.g., customers' information, business secrets, personal communication). Furthermore, the available intrusion detection datasets are not often all-inclusive, and the flow samples are not complete to cover all traffic types. The CICDDoS2019 is one of the up-to-date datasets including different DDoS attack types, which could be done in the application layer on top of *TCP/UDP*. CICDDoS2019 classifies DDoS attacks into two taxonomies. The first one is exploitation-based attacks, and the second one is reflection-based attacks. Two days were spent gathering all traffic information for the training phase and test phase. The first day was dedicated for the training on 12th of January 2019 and contained 12 types of DDoS attacks as follows: SYN, WebDDoS, MSSQL, UDP,

---

[2] https://www.unb.ca/cic/datasets/ddos-2019.html.

SNMP, NetBIOS, DNS, LDAP, TFTP, NTP, UDP-Lag, and SSDP DDoS based attack. On the second day, March 11, 2019, the testing data ware collected and included seven DDoS attacks: SYN, MSSQL, LDAP, UDP, Port Scan, UDP-Lag, and NetBIOS. In the CICDDoS2019 dataset, each flow traffic has more than 80 features, which were extracted by CICFlowMeter tools [2]. Further, the dataset is open and accessible in two format types (PCAP file and CSV file) on the website of the Canadian Institute for Cyber Security [1,22].

4.2 Dataset Manipulation

The CICDDoS2019 contains some undesired attributes or errors that do not fit deep learning algorithms. Therefore, it is mandatory to first prepare the dataset for the 1D-CNN model. As mentioned above, CICDDoS2019 includes more than 80 features, which are irrelevant for some attack types and relevant for other types. Hence, the following steps are implemented on the dataset before being fed to the model in this work.

1. All irrelevant features are removed, such as source and destination IP, source and destination port, timestamp, and flow ID. They do not introduce any valuable pointer to any type of attack and are different from one network to another. Moreover, the adversary usually crafts the same normal user address information. As a result, training the 1D CNN model with socket information would present an overfitting problem. In the end, 77 features are prepared as the model input after removing these unwanted features.
2. In the second step, all traffic flows show errors in the CICDDoS2019 dataset. These errors (e.g., non-value features or infinity values) are removed.
3. All features in the traffic have specific numerical values, which might lead to a wrong classification if the 1D-CNN model uses them as they are. Therefore, all feature values are normalized using the next equation:

$$V_{new} = \frac{V_{old} - MIN}{MAX - MIN} \tag{1}$$

Where **V**: the new and the old value, **MIN**: the minimum value for all considered values of the same feature in the dataset, **MAX**: the maximum value for all considered values of the same feature in the dataset.
4. In the end, the symbolic feature of the attack label is encoding by 0 for all the DDoS attack traffic flows and 1 for the benign traffic flows to be ready for training the suggested binary classification model.

4.3 Proposed 1D-CNN Architecture

The proposed 1D-CNN architecture model has two convolutional layers with 32 and 16 filters, where each filter has a size of 1×3 and a stride of 1. Each convolutional layer is followed by a max-pooling layer with a size of 1×2. The 'same' padding is used in the convolutional layers to make the output the same as the input. Thus, the input gets fully covered by the filter and the specified stride. Two fully connected layers are used with some units equal to 16 or 8, respectively. The final layer is the sigmoid layer to classify the network traffic into normal (0) or malicious traffic (1). Furthermore, a Nadam optimizer and a mean squared error (MSE) are used for the model. The hyper-parameter configuration is 100, 10, and 0.001 for the batch size, epoch, and learning rate. At the same time, the binary cross-entropy is used as a loss function in the output layer. The non-linear ReLU (rectified linear unit) activation function is used for all layers except the last layer.

To investigate the suitability for the SDN environment, the 1D-CNN model is applied as a unit of the controller. However, using the model for every packet or traffic is not acceptable because of the high computational complexity [25]. Therefore, three units are required before the controller decides to mitigate any malicious source:

CICFlowMeter Unit In this paper, the CICFlowMeter version 3 [2] tool is used to capture all bidirectional packets of the SDN controller port number 6653. CICFlowMeter is a network tool that analyzes all bidirectional packets to generate specific information about the network traffic flow for a certain port, device, or any network application. CICFlowMeter defines the first packet no matter whether it is in the forward (source to destination) or backward (destination to source) direction. It provides statistical information in a CSV format file with more than 80 features for each network traffic such as duration, length of packets, number of packets, number of bytes, etc. [1,20]. Hence, CICFlowMeter prepares the required information in a CSV file.

Entropy Information Unit monitors the randomness of the *Packet_In* messages. In the injection attack, the adversary tries to flood the SDN controller or the network with false header packet information (e.g., source or destination addresses). Once the attack is launched, the *Miss Table* entry of the connected switch encapsulates the packet as *Packet_In* message and submits it through the OpenFlow channel to the SDN controller. As a result, the value of randomness would be higher than the predefined threshold, which provides a sign for suspicious behavior of the input port to the SDN controller. However, a flash crowd situation mostly has the same behavior as malicious packets (e.g., packets per second, bits per second, and flows per second), and hence the entropy information is not sufficient to identify whether there is an attack or not [12,25]. The entropy calculation is real-time capable and treats numerous traffic flows with a low computing overhead [27]. Once the SDN controller receives the *Packet_In* message, the sensitive features are extracted and stored in the entropy window buffer for randomness calculation. The 1D-CNN unit is transferred from silent into the active state if the entropy threshold is broken. Generally, two components are essential to detect any abnormal behavior for DDoS using entropy, (a) the window size and (b) the entropy threshold value. The entropy window contains the sensitive features for all received *Packet_In* messages and the in-port information of the connected switch. Thereby, the entropy value is immediately calculated as long as the entropy window is full with a specified number of packets (in this work 100). The Entropy Information Unit counts the number of the repeated appearances for every single sensitive feature ϕ. It then calculates the probability of occurrence $P(\phi)$ based on the entropy window size W, like the following:

$$P(\phi) = \phi/W \tag{2}$$

and the Entropy value E of the full window is calculated in Eq. 3:

$$E = \sum -P(\phi)log_2 P(\phi) \tag{3}$$

In other words, once the entropy threshold is violated, and the 1D-CNN finds an attack during the last 50 flows information in the CSV file, the controller determines the malicious ports by calculating the probability of the port $(P(\phi))$ in the entropy window. Hence, the port of the highest $P(\phi)$ is blocked. The entropy value is recalculated by dividing the blocked port probability by entropy slots in the threshold window. If the entropy value is less than the threshold or equal, the block port process stops. Otherwise, the controller will block the port of the second weight and so on.

Dynamic Threshold Unit calculates the dynamic threshold, considering that the network traffic characteristics are changeable and do not have the same conduct. Therefore, the value of the entropy threshold should be adaptive. According to Algorithm 1, we have another window (threshold window) whose size is changeable. In case there is no suspicious *Packet_In* traffic, the controller inserts the new entropy value into the threshold window. It then obtains the variance (V), which is twice calculated, firstly without the newly inserted entropy value (old_V), and secondly without the oldest entropy value in the threshold window (new_V), as defined by Eq. 4 where n is the number of considered entropy values:

$$V = \frac{\sum_{i=1}^{n}(E_i - M)^2}{n} \tag{4}$$

As a rule, a small variance value indicates that the newly inserted entropy is close to the mean (M). On the contrary, a high variance value indicates that the new entropy value passes away from the mean. The quantity (Q) is the largest variance inside of a subset of all entropy values. Q aims to quantify the maximum V in the threshold window according to the next equation:

$$Q = \frac{(M - S)(G - M)}{V} \tag{5}$$

S is the smallest entropy value, and G is the greatest entropy value in the threshold window. Besides, the direct and inverse values are obtained to calculate the $DiffRatio$ of the change. $Direct$ measures the mean changes between

the current and last windows, and $Inverse$ represents its inverse and measures
the mean changes between the last and current windows.

$$Direct = new\_V/old\_V \tag{6}$$

$$Inverse = old\_V/new\_V \tag{7}$$

$$DiffRatio = \sqrt{(direct - inverse)^2} \tag{8}$$

Once $DiffRatio$ is greater than $(1+\beta)$ and the $oldQ$ greater than $newQ$, the
algorithm increases the window size by rounding the quantity Q value to the
nearest integer. Nevertheless, when $DiffRatio$ is greater than $(1 + \beta)$ and
$oldQ$ is less than $newQ$, the algorithm reduces the window size by rounding
the quantity value to the nearest integer. β cancels the superfluous overhead
and equals 5%, representing a 95% confidence interval.

Algorithm 1: Dynamic Window Size

Input:

- Mean of the current window values (M_{new}),
- Mean of the previous window values (M_{old}),
- Current dynamic window size (CWsize), smallest value (S), greatest value (G)

Result: Next Window Size (NWsize)
initialization;
$Variance_{new} = \frac{\sum_{i=1}^{n}(value_i - M_{new})^2}{n}$
$Variance_{old} = \frac{\sum_{i=1}^{n}(value_i - M_{old})^2}{n}$
$Quantity = (M_{new} - S) * (G - M_{new})/Variance_{new}$
$Direct = Variance_{new}/Variance_{old}$
$Inverse = Variance_{old}/Variance_{new}$
$DiffRatio = \sqrt{(direct - inverse)^2}$
if $DiffRatio > 1 + \beta$ **then**
    if $Variance_{new} > Variance_{old}$ **then**
       | $NWSize = CWsize + \lfloor Quantity \rceil$
    else
       | $NWSize = CWsize - \lfloor Quantity \rceil$
    end
end
return NWsize

As is depicted in Fig. 2, the unit uses the standard normal distribution to
calculate the entropy threshold. The work follows the coverage theorem of normal
distribution (what is well known as 68 – 95 – 99.7 (empirical) rule or the 3-sigma
rule) [12]. The standard deviation (σ) of the mean represents 68% of the normal
distribution values in the area between $\sigma - \mu$ and $\sigma - \mu$. Also, 95% of the values
are in the area between $2\sigma - \mu$ and $2\sigma - \mu$. However, 99.7% of the values lie

in the area between $3\sigma - \mu$ and $3\sigma - \mu$. Therefore, the SDN controller has all entropy values in the threshold window after a specific time. The unit calculates the mean of the threshold window (μ) to obtain the value of the σ according to Eq. 9. In the end, the next point in the threshold is calculated by Eq. 10.

$$\sigma = \sqrt{\frac{\sum (E - \mu)^2}{CW size}} \tag{9}$$

$$Threshold = \mu - 3\sigma \tag{10}$$

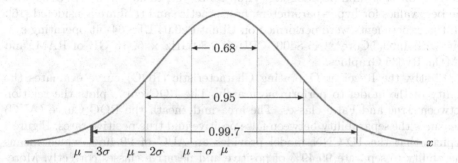

Fig. 2. 3-Sigma rule

5   Evaluation and Analysis

This section introduces the details of the experimental setup to analyze the performance of the proposed method. The evaluation is separated into two subsections. In Subsect. 5.1, the experimental results of the proposed model (1D-CNN) are discussed. In Subsect. 5.2, the implementation of the model is shown based on the SDN environment.

5.1   1D-CNN Evaluation Criteria

The four metrics accuracy, precision, recall, and F1 score are used to have a systematic benchmarking analysis with other related approaches and evaluate the 1D-CNN model. The mathematical representation of these metrics are based on the following equations:

$$Accuracy = \frac{TP + TN}{TP + TN + FP + FN} \tag{11}$$

$$Precision = \frac{TP}{TP + FP} \tag{12}$$

$$Recall = \frac{TP}{TP + FN} \qquad (13)$$

$$F1\ Score = \frac{2 \cdot Precision \cdot Recall}{Precision + Recall} \qquad (14)$$

Here, True Positive (TP) and True Negative (TN) represent the correctly predicted values. In contrast, False Positives (FP) and False Negatives (FN) are wrongly classified events. The experiments use four metrics to compare the proposed model against the four machine learning models (ID3, random forest, Naïve Bayes, and logistic regression) proposed [22] with CICDDoS2019 for performance validation. TensorFlow and Keras framework construct all models in the Python programming language [4,6]. Moreover, to construct 1D-CNN with the best values for hyper-parameters, best practice and trial are considered [16]. All the experiments were performed on Ubuntu 20.04 LTS 64-bit operating system with Intel®Core™ i5-8400, CPU @ 2.80 GHz × 6, 16 GB of RAM, and AMD®Rv635 Graphics.

Firstly, the Receiver Operating Characteristic (ROC) curve evaluates the ability of the model to perform accurately. The ROC curve plots the relation between True and False classes. The area underneath the ROC Curve (AUC) measures the separability between false positive and true positive rates. Figure 3 depicts that the 1D-CNN model provides an AUC of 99.49%, which means the ability to separate 99.49% of positive and negative classes correctly. Moreover, the 1D-CNN model introduces a notable enhancement compared to other machine learning models. 1D-CNN provides 12% 22%, 10%, and 12% enhancement compared to Random Forest, Logistic regression, Naïve Bayes, and ID3, respectively.

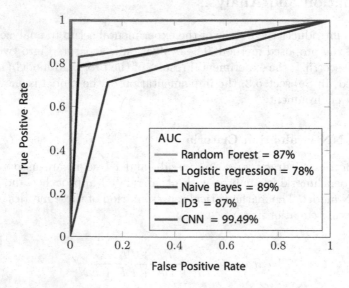

Fig. 3. Receiver Operating Characteristic (ROC)

Table 1. Detection performance comparison

Learning model	Accuracy	Precision	Recall	F1 score
Random Forest	0.90	0.92	0.78	0.84
Logistic Regression	0.80	0.73	0.72	0.72
Naïve Bayes	0.91	0.92	0.82	0.87
ID3	0.90	0.92	0.78	0.85
1D-CNN	**0.9945**	**0.9884**	**0.9962**	**0.9923**

Table 1 shows more details about the superiority of the 1D-CNN model over the other models. The enhancements are presented in terms of accuracy, precision, recall, and F1 score. 1D-CNN achieves better results with around 10% to 20% accuracy enhancements compared to the other models. In addition, our model shows best results in precision with 6% to 25% enhancements, in terms of recall with 17% to 27%, and 12% to 27% in terms of F1 score. Table 2 illustrates the confusion matrix information to describe the classification performance of 1D-CNN. The table shows the correct and false predictions. It makes a comparison that indicates that the 1D-CNN model outperforms all other machine learning models in terms of four different events (TP, FP, TN, FN) obtained using our proposed approach.

Table 2. Confusion matrix.

Learning model	TN	FP	FN	TP
Random Forest	0.96	0.04	0.21	0.79
Logistic Regression	0.86	0.14	0.28	0.72
Naïve Bayes	0.96	0.04	0.17	0.83
ID3	0.97	0.03	0.21	0.79
1D-CNN	**0.99**	**0.006**	**0.003**	**0.99**

5.2   Effectiveness for Applying 1D-CNN in SDN

Mininet is used to evaluate a small network with two switches (S1, S2) managed by Ryu Controller [3] to investigate the performance of the proposed 1D-CNN model based on entropy information. S1 connects to four normal users and four attackers. S2 connects to four servers. The normal users use Scapy [5] to read benign UDP PCAP files, which have numbers from 500 to 749 in the CICD-DoS2019 dataset, and send these packets to the servers. After a short time, the attackers also start reading the malicious UDP PCAP files, which have numbers from 750 to 818 in CICDDoS2019, spoofing the MAC address for every packet.

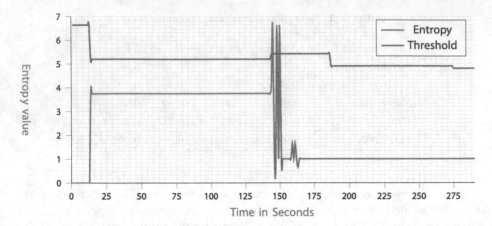

Fig. 4. Destination entropy value

The simulation was executed twice to check out the entropy behavior for crafted packets with MAC source or destination in both cases.

Figures 5 and 4 depict the behavior of the entropy values during the simulation time, in the case of the attacker crafting MAC source address or destination address. In the beginning, the threshold value started from the topmost value for log_2 of the window size (100), which is equal to 6.643. Once the packets start entering the network, the threshold will be changing because of the new *Packet_In* messages. The simulation monitors the last entropy value, so when the users' communications do not need to request the controller for intervention, the entropy value will not change. This is the case from second 19 until the beginning of the attack around 140 and after the controller stops the attack until the end. The entropy value crosses the threshold more than once in both figures because the attacker started the attack after reading the PCAP files, which launches the attack at different times. However, the features could be changing, where the switch table identifies the packet according to the used SDN controller instructions by many different features such as IP addresses, MAC addresses, protocol, output or input port, VLAN, MPLS, etc. [27].

Figure 6 depicts the behavior of the CPU usage in three states: (1) normal without DDoS attack, (2) attack state without defense mechanism, and (3) with 1D-CNN controller structure based on entropy information. In the normal state, the CPU usage rate is high between seconds 0 and 25 because normal traffic is sent through the network for the first time, and a relatively large number of *Packet_In* messages was triggered to establish flow entries. In the attack state, a very high rate of CPU usage is notable when the simulation launched the malicious traffic (after around 135 s) in addition to the high rate in the beginning. However, this does not occur with our proposal; the CPU usage is high for the beginning only, increasing a little bit before the controller blocks the connected ports. The controller needs a short time to read the last 100 traffic flows saved as a CSV file to verify any attack towards the controller during the past time.

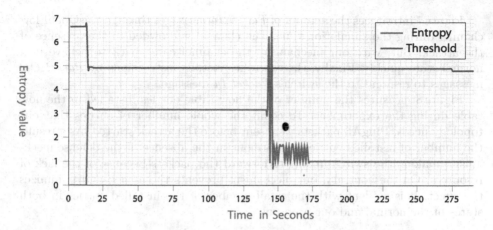

Fig. 5. Source entropy value

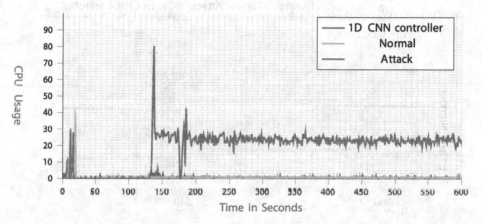

Fig. 6. CPU usage

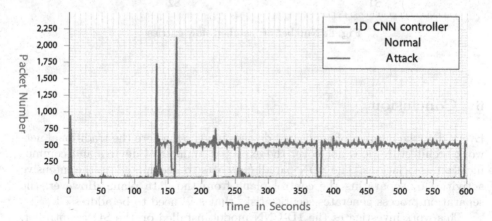

Fig. 7. OpenFlow channel bandwidth

Figure 7 introduces the amount of *Packet_In* messages through the OpenFlow channel during the simulation time. The channel is crowded in the presence of the attack. But, with our mechanism, the channel came back to behave in a normal state quickly. Finally, the controller installs instructions by *Packet_Out* messages to respond to the switch's *Packet_In* message.

Figure 8 measures the number of the flow table entries installed in the flow table during the experiment. To count the whole number of entries that our topology needs, PingAll packets were sent before the attack started. As a result, the number of installed entries is enormous in the absence of the defense mechanism, making the switch unable to forward the packets because of the lack of resources. On the contrary, our mechanism prevents the attack and minimizes the effect. It is evident with the small number of the installed entries in both states of the normal and our proposal.

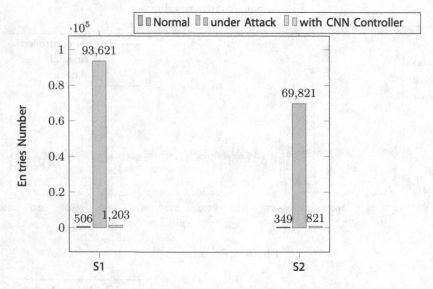

Fig. 8. Number of installed flow entries

6   Conclusion

Every day, the number of connected devices increases where the traditional network architecture does not cover the expected number of devices and communication associations. The SDN paradigm starts to accommodate the intensive activities by separating the control plane from the data plane. However, the separation process generates security risks that still need to be addressed.

This work investigates the 1D-CNN model installed on the SDN controller. The model is compared to other state-of-the-art models. Besides, the 1D-CNN is proposed based on entropy information as IDS to tackle the security issues of the SDN controller. The simulations prove that the proposed model shows many advantages compared to state of the art.

As future work, 1D-CNN should be improved by proposing different techniques to enhance its ability for anomaly detection. The work shall also investigate the implementation of very deep learning algorithms to improve the performance of the CNN for IDS. Finally, we will utilize other datasets to prove the performance of our approach.

References

1. Canadian Institute for Cybersecurity. https://www.unb.ca/cic/. Accessed 01 June 2021
2. CICFlowMeter V3 Python Implementation. https://pypi.org/project/cicflowmeter/. Accessed 01 June 2021
3. Component-Based Software Defined Networking Framework - Build SDN Agilely. https://ryu-sdn.org/. Accessed 01 June 2021
4. Keras Framework. https://keras.io/. Accessed 01 June 2021
5. Packet crafting for Python2 and Python3. https://scapy.net/. Accessed 01 June 2021
6. Abadi, M., et al.: TensorFlow: large-scale machine learning on heterogeneous systems (2015). https://arxiv.org/abs/1603.04467
7. Ahuja, N., Singal, G., Mukhopadhyay, D.. DLSDN. deep learning for DDOS attack detection in software defined networking. In: 2021 11th International Conference on Cloud Computing, Data Science & Engineering (Confluence), pp. 683 688. IEEE (2021)
8. Al-Qatf, M., Lasheng, Y., Al-Habib, M., Al-Sabahi, K.: Deep learning approach combining sparse autoencoder with SVM for network intrusion detection. IEEE Access 6, 52843–52856 (2018)
9. Alkasassbeh, M., Almseidin, M.: Machine learning methods for network intrusion detection. arXiv preprint arXiv:1809.02610 (2018)
10. Althubiti, S.A., Jones, E.M., Roy, K.: LSTM for anomaly-based network intrusion detection. In: 2018 28th International Telecommunication Networks and Applications Conference (ITNAC), pp. 1–3. IEEE (2018)
11. Assis, M.V., Carvalho, L.F., Lloret, J., Proença, M.L., Jr.: A GRU deep learning system against attacks in software defined networks. J. Netw. Comput. Appl. 177, 102942 (2021)
12. Conti, M., Lal, C., Mohammadi, R., Rawat, U.: Lightweight solutions to counter DDoS attacks in software defined networking. Wireless Netw. 25(5), 2751–2768 (2019)
13. Dehkordi, A.B., Soltanaghaei, M., Boroujeni, F.Z.: The DDoS attacks detection through machine learning and statistical methods in SDN. J. Supercomput. 77(3), 2383–2415 (2021)
14. Deng, S., Gao, X., Lu, Z., Gao, X.: Packet injection attack and its defense in software-defined networks. IEEE Trans. Inf. Forensics Secur. 13(3), 695–705 (2017)
15. Elsayed, M.S., Jahromi, H.Z., Nazir, M.M., Jurcut, A.D.: The role of CNN for intrusion detection systems: an improved CNN learning approach for SDNs. In: Perakovic, D., Knapcikova, L. (eds.) FABULOUS 2021. LNICSSITE, vol. 382, pp. 91–104. Springer, Cham (2021). https://doi.org/10.1007/978-3-030-78459-1_7

16. Elsayed, M.S., Le-Khac, N.A., Jurcut, A.D.: InSDN: a novel SDN intrusion dataset. IEEE Access **8**, 165263–165284 (2020)
17. Hu, Z., Wang, L., Qi, L., Li, Y., Yang, W.: A novel wireless network intrusion detection method based on adaptive synthetic sampling and an improved convolutional neural network. IEEE Access **8**, 195741–195751 (2020)
18. Karan, B., Narayan, D., Hiremath, P.: Detection of DDoS attacks in software defined networks. In: 2018 3rd International Conference on Computational Systems and Information Technology for Sustainable Solutions (CSITSS), pp. 265–270. IEEE (2018)
19. Krishnan, P., Duttagupta, S., Achuthan, K.: VARMAN: multi-plane security framework for software defined networks. Comput. Commun. **148**, 215–239 (2019)
20. Lashkari, A.H., Draper-Gil, G., Mamun, M.S.I., Ghorbani, A.A.: Characterization of tor traffic using time based features. In: International Conference on Information Systems Security and Privacy (ICISSP 2017), pp. 253–262 (2017)
21. Prasath, M.K., Perumal, B.: A meta-heuristic Bayesian network classification for intrusion detection. Int. J. Network Manage **29**(3), e2047 (2019)
22. Sharafaldin, I., Lashkari, A.H., Hakak, S., Ghorbani, A.A.: Developing realistic distributed denial of service (DDoS) attack dataset and taxonomy. In: 2019 International Carnahan Conference on Security Technology (ICCST), pp. 1–8. IEEE (2019)
23. Swami, R., Dave, M., Ranga, V.: Defending DDoS against software defined networks using entropy. In: 2019 4th International Conference on Internet of Things: Smart Innovation and Usages (IoT-SIU), pp. 1–5. IEEE (2019)
24. Tang, T.A., Mhamdi, L., McLernon, D., Zaidi, S.A.R., Ghogho, M.: Deep recurrent neural network for intrusion detection in SDN-based networks. In: 2018 4th IEEE Conference on Network Softwarization and Workshops (NetSoft), pp. 202–206. IEEE (2018)
25. Wang, L., Liu, Y.: A DDoS attack detection method based on information entropy and deep learning in SDN. In: 2020 IEEE 4th Information Technology, Networking, Electronic and Automation Control Conference (ITNEC), vol. 1, pp. 1084–1088. IEEE (2020)
26. Xiao, Y., Xing, C., Zhang, T., Zhao, Z.: An intrusion detection model based on feature reduction and convolutional neural networks. IEEE Access **7**, 42210–42219 (2019)
27. Xu, J., Wang, L., Xu, Z.: An enhanced saturation attack and its mitigation mechanism in software-defined networking. Comput. Netw. **169**, 107092 (2020)
28. Yamashita, R., Nishio, M., Do, R.K.G., Togashi, K.: Convolutional neural networks: an overview and application in radiology. Insights Imaging **9**(4), 611–629 (2018)
29. Ye, J., Cheng, X., Zhu, J., Feng, L., Song, L.: A DDoS attack detection method based on SVM in software defined network. Secur. Commun. Netw. **2018** (2018)

Multi-Armed Bandit-Based Channel Hopping: Implementation on Embedded Devices

Konrad-Felix Krentz$^{(\boxtimes)}$, Alex Kangas, and Thiemo Voigt

Uppsala universitet, P.O. Box 256, 751 05 Uppsala, Sweden
konrad.krentz@it.uu.se, alex.kangas.5644@student.uu.se,
thiemo.voigt@it.uu.se

Abstract. Simulations have shown multi-armed bandit (MAB) algorithms to be suitable for optimizing channel hopping in IEEE 802.15.4 networks. Thus far, however, there appears to be no practical implementation of this approach, presumably because typical IEEE 802.15.4 nodes lack both floating-point unit (FPUs) and big amounts of random-access memory (RAM). In this paper, we propose fixed-point arithmetic and implementation shortcuts to circumvent these constraints. We focus on two specific multi-armed bandit (MAB) algorithms, namely sliding-window upper confidence bound (SW-UCB) and its predecessor discounted UCB (D-UCB). SW-UCB is particularly promising since it requires only tractable fixed-point arithmetic, while yielding high packet delivery ratios (PDRs) according to prior work. D-UCB, on the other hand, additionally opens up an implementation shortcut that saves RAM. Our implementations of SW-UCB and D-UCB are integrated into Contiki-NG, yet can also be used out-of-tree in a simulation environment. We show our SW-UCB (resp. D-UCB) implementation to attain PDRs of 98.6% (resp. 99.2%) under appropriate parameter settings in the context of intra-body communication. Also, we demonstrate D-UCB to incur a moderate RAM, program memory, and processing overhead on CC2538 SoCs, whereas we find SW-UCB too RAM-consuming for these chips. Finally, using Monte Carlo simulations, we show our SW-UCB and D-UCB implementations to perform equally well as floating-point counterparts.

1 Introduction

Early efforts on applying channel hopping to IEEE 802.15.4 networks followed various goals. Two initial ones were to increase throughput and to lower latencies through the use of additional channels [8, 20, 22, 33]. Further ones were to evade channels that are subject to external interference, destructive fading effects, or jamming attacks [8, 32, 37]. Lastly, there was also the goal of extending communication range since particular channels may benefit from constructive fading effects [32]. Meanwhile, channel hopping became an integral part of the IEEE 802.15.4 standard [2]. The 2020 version of IEEE 802.15.4 specifies three medium

© Springer Nature Switzerland AG 2022
É. Renault et al. (Eds.): MLN 2021, LNCS 13175, pp. 29–47, 2022.
https://doi.org/10.1007/978-3-030-98978-1_3

access control (MAC) protocols that support channel hopping, namely coordinated sampled listening (CSL), time-slotted channel hopping (TSCH), and carrier sense multiple access with collision avoidance (CSMA-CA).

Current research seeks to advance the now standardized blind channel hopping, where each node follows a fixed channel hopping pattern, towards an adaptive channel hopping, where each node selects channels based on their performance [11–15,17,18,21,27,34,41]. One way to implement adaptive channel hopping is to let one or more nodes assess the available channels and subsequently blacklist bad channels (or whitelist good channels) either cluster- or network-wide [12–15,21,27,34,41]. Yet, this approach makes the simplifying assumption that all channels work equally well between all pairs of nodes in a cluster or network. In general, this is not the case due to the varying exposure to interference, as well as the spatial dependence of fading effects. Moreover, channel blacklisting (resp. whitelisting) typically incurs a communication overhead for synchronizing the blacklisted (resp. whitelisted) channels [14]. Alternatively, there are two ways to implement adaptive channel hopping in a localized manner. First, each pair of nodes may negotiate its own blacklist or whitelist of channels [17,18]. While this avoids assumptions on the radio environment, a communication overhead persists. Second, each node can restrict itself to its local information, e.g., acknowledged and unacknowledged radio transmissions, to learn which channels work well for communication with a particular neighbor [11]. This learning process can be efficiently guided by a multi-armed bandit (MAB) algorithm [11]. Such algorithms find trade-offs between exploitation and exploration, which is exactly the problem that arises in self-adaptive channel hopping.

Despite the advantages of MAB-based channel hopping over both channel blacklisting and whitelisting, no practical implementations seem to exist. Previous work employing MAB algorithms in IEEE 802.15.4 networks focused on simulations and neglected practical implementation issues [11,15,21]. Implementing MAB algorithms on typical IEEE 802.15.4 nodes is, however, challenging. For one thing, IEEE 802.15.4 nodes often lack a floating-point unit (FPU). For another, most IEEE 802.15.4 nodes are severely constrained in terms of random-access memory (RAM). For example, a CC2538 system on chip (SoC) only has up to 32 KB of RAM, half of which is only usable for stack data [35].

This paper makes two main contributions:

- We propose fixed-point arithmetic and implementation shortcuts to practically implement sliding-window upper confidence bound (SW-UCB) and discounted UCB (D-UCB) on resource-constrained IEEE 802.15.4 nodes. Our choice of SW-UCB is motivated by Dakdouk et al.'s finding that SW-UCB performed just slightly worse than the best considered MAB algorithm, which was an advanced version of Thompson Sampling [11]. Yet, Thompson Sampling involves drawing samples from a beta distribution, necessitating multiple invocations of the gamma function. SW-UCB, by contrast, is much more tractable in terms of fixed-point arithmetic. Furthermore, D-UCB, a predecessor of SW-UCB, scales better in terms of RAM than SW-UCB.

- We give an extensive experimental evaluation of our SW-UCB and D-UCB implementations, comprising Monte Carlo simulations, measurements of their attained real-world packet delivery ratios (PDRs), as well as measurements of their RAM, program memory, and processing overhead on CC2538 SoCs.

The rest of this paper is organized as follows. Section 2 introduces SW-UCB and D-UCB. Section 3 complements our discussion of related work. Section 4 presents our proposed fixed-point implementations of SW-UCB and D-UCB, points out implementation shortcuts, as well as describes the integration of SW-UCB and D-UCB into IEEE 802.15.4 CSL. Section 5 gives the experimental evaluation. Finally, Sect. 6 concludes and suggests directions for future work.

2   Background

Both SW-UCB and D-UCB solve the piecewise-stationary MAB problem, which may be stated as follows [39]. An agent pulls one of K independent arms at each of T time steps. After pulling an arm $k \in \{1, \ldots, K\} = [K]$ at time step $t \in \{1, \ldots, T\} = [T]$, the agent gets a reward $x \in [0,1]$. x is generated by the random variable $X_{k,t}$ that produces rewards for pulling k at time t. The degree of nonstationarity is measured by the number of segments $\varUpsilon$:

$$\varUpsilon = 1 + \sum_{t=1}^{T-1} \mathbb{1}_{\{\mathbb{E}[X_{k,t}] \neq \mathbb{E}[X_{k,t+1}] \text{ for some } k \in [K]\}}, \tag{1}$$

where

$$\mathbb{1}_{\{c\}} = \begin{cases} 1, & \text{if the condition } c \text{ is met} \\ 0, & \text{otherwise} \end{cases} \tag{2}$$

The agent is neither given a priori information about the time steps when the expected reward of one or more arms changes, nor about the number of segments $\varUpsilon$, nor about the probability distributions of the rewards.

Both SW-UCB and D-UCB tailor the so-called UCB1 algorithm to piecewise-stationary environments. The UCB1 algorithm is listed as Algorithm 1 [5]. UCB1 itself is designed for the special case that $\varUpsilon = 1$, i.e., a stationary environment. Initially, UCB1 samples each arm once. At Line 6, UCB1 then uses Hoeffdings inequality, from which it follows that for arm $k \in [K]$ at time step $t \in [T]$:

$$\mathbb{P}\left[\mathbb{E}[X_{k,t}] > \frac{1}{n_k} \sum_{s=1}^{t-1} \mathbb{1}_{\{I_s = k\}} x_{k,s} + \sqrt{\frac{2\ln(t-1)}{n_k}}\right] \leq (t-1)^{-4} \tag{3}$$

By pulling the arm with the highest upper confidence bound, UCB1 initially tends to pull arms for which less historical data is available. Besides, the time-dependent confidence level $1 - (t-1)^{-4}$ widens the upper confidence bounds over time, thus inclining UCB1 to continuously explore less rewarding arms, too.

SW-UCB and D-UCB tailor UCB1 to piecewise-stationary environments ($\varUpsilon \geq 1$) by weighing more on recent awards [16,23]. The approach of SW-UCB

Algorithm 1. UCB1

Require: T: time horizon T, K: number of arms
 1: **for** $t = 1$ **to** T **do**
 2:    **if** $t \leq K$ **then**
 3:        $I_t \leftarrow t$
 4:    **else**
 5:        $n_k \leftarrow \sum_{s=1}^{t-1} \mathbb{1}_{\{I_s=k\}}$, for all $k \in [K]$
 6:        $I_t \leftarrow \arg\max_{k \in [K]} \frac{1}{n_k} \sum_{s=1}^{t-1} x_{k,s} + \sqrt{\frac{2\ln(t-1)}{n_k}}$
 7:    **end if**
 8:    Pull arm I_t and receive reward $x_{I_t,t} \leftarrow X_{I_t,t}$
 9:    $x_{k,t} \leftarrow 0$, for all $k \in [K] \setminus \{I_t\}$
10: **end for**

Algorithm 2. SW-UCB

Require: T: time horizon T, K: number of arms,
    $\xi > 0$: exploration tendency,
    $\tau > K$: length of sliding window
 1: **for** $t = 1$ **to** T **do**
 2:    **if** $t \leq K$ **then**
 3:        $I_t \leftarrow t$
 4:    **else**
 5:        $n_k \leftarrow \sum_{s=\max\{1,t-\tau\}}^{t-1} \mathbb{1}_{\{I_s=k\}}$, for all $k \in [K]$
 6:        **if** $n_k = 0$ for any $k \in [K]$ **then**
 7:            $I_t \leftarrow k$
 8:        **else**
 9:            $I_t \leftarrow \arg\max_{k \in [K]} \frac{1}{n_k} \sum_{s=\max\{1,t-\tau\}}^{t-1} x_{k,s} + \sqrt{\frac{\xi \ln \min\{t-1,\tau\}}{n_k}}$
10:        **end if**
11:    **end if**
12:    Pull arm I_t and receive reward $x_{I_t,t} \leftarrow X_{I_t,t}$
13:    $x_{k,t} \leftarrow 0$, for all $k \in [K] \setminus \{I_t\}$
14: **end for**

Algorithm 3. D-UCB

Require: T: time horizon T, K: number of arms,
    $\xi > 0$: exploration tendency,
    $\gamma \in (0, 1)$: discount factor
 1: **for** $t = 1$ **to** T **do**
 2:    **if** $t \leq K$ **then**
 3:        $I_t \leftarrow t$
 4:    **else**
 5:        $n_k \leftarrow \sum_{s=1}^{t-1} \gamma^{t-s-1} \mathbb{1}_{\{I_s=k\}}$, for all $k \in [K]$
 6:        $I_t \leftarrow \arg\max_{k \in [K]} \frac{1}{n_k} \sum_{s=1}^{t-1} \gamma^{t-s-1} x_{k,s} + \sqrt{\frac{\xi \ln \sum_{l=1}^{K} n_l}{n_k}}$
 7:    **end if**
 8:    Pull arm I_t and receive reward $x_{I_t,t} \leftarrow X_{I_t,t}$
 9:    $x_{k,t} \leftarrow 0$, for all $k \in [K] \setminus \{I_t\}$
10: **end for**

---

is to only keep a sliding window of the most recent rewards. SW-UCB is listed as Algorithm 2. D-UCB is listed as Algorithm 3. Its approach is to use a discount factor γ so that old rewards diminish over time.

3   Further MAB Algorithms

In the context of channel hopping, it is crucial to adopt MAB algorithms that can handle changing environments. In the literature, there emerged two ways to adapt to changing environments, namely the passively and the actively adaptive policies [28]. The approach of the passively adaptive policies is to weigh more on recent awards than on old rewards [6,10,16,23,31]. For example, SW-UCB and D-UCB belong to the passively adaptive policies [16,23]. The approach of the actively adaptive policies, on the other hand, is to actively detect change points and to restart the learning process when detecting a change point [9,19,28–30]. Unfortunately, while actively adaptive policies performed particularly well in Dakdouk et al.'s experiments [11], they incur a high RAM and processing overhead, rendering them unsuitable for constrained IEEE 802.15.4 nodes.

Another categorization of MAB algorithms is based on which level of non-stationary they are designed for [4]. Basic MAB algorithms are only designed for a stationary environment. At the other extreme, Auer et al. considered an environment, where rewards are chosen by an adversary to encompass any kind of non-stationarity [6]. Between both extremes are the MAB algorithms for piecewise-stationary and drifting environments. The piecewise-stationary environment described in the previous section is due to Yu et al. [39]. Besides, Besbes et al. introduced a drifting environment, where the expected rewards may only change within the confines of a variation budget [7].

Reinforcement learning (RL) addresses more general problems than MAB algorithms. One main generalization of RL problems compared to MAB problems is that actions in RL problems may not only influence immediate rewards, but also future rewards [40]. However, this generalization is not normally required in the context of channel hopping as the choice for a channel normally has no controllable effect on future channel performances. A second main generalization is that the agent in an RL problem may base its decisions not only on past rewards, but also on contextual information about the environmental state. This second generalization appears more useful in the context of channel hopping as this, e.g., allows an agent to take mobility information into account. Accordingly, a promising hybrid between the RL and the MAB problem is the contextual MAB problem [40]. In this variant, the agent may take contextual information into account, while avoiding the complexity of the full RL problem. That said, the additional complexity entailed by contextual MAB algorithms still mismatches the constrained hardware of typical IEEE 802.15.4 nodes.

4   SW-UCB and D-UCB Under Constraints

In this section, we detail the fixed-point arithmetic and implementation short-cuts we used to realize SW-UCB and D-UCB on our target hardware platform, namely the CC2538 SoC [35]. Furthermore, we discuss our integration of SW-UCB and D-UCB into the IEEE 802.15.4 CSL MAC protocol.

4.1 Relevant Fixed-Point Arithmetic

Base-2 fixed-point arithmetic enables representing a fractional number as an ordinary n-bit integer in memory. The numerical value of such an n-bit integer a is $a \times 2^x$, where the negative integer x specifies the position of the radix point. Usually, x and n are program-wide constants, which allow for the required length of both characteristics and mantissas. For example, we used $n = 32$ and $x = -16$ in our fixed-point implementation of SW-UCB, whereas we used $n = 32$ and $x = -22$ in our fixed-point implementation of D-UCB.

Subtracting and adding two base-2 fixed-point numbers a, b can be efficiently implemented through a single integer subtraction or addition, respectively:

$$a \times 2^x \pm b \times 2^x = (a \pm b) \times 2^x \tag{4}$$

Multiplication and division is more involved. When multiplying two fixed-points numbers, each number's scaling factor will also be multiplied. For instance, if two base-2 fixed-point numbers both have a scaling factor 2^x, then their product will have a scaling factor of 2^{2x}. The original fixed-point format can be restored via bit shifting the product by $|x|$ to the right. Similarly, the scaling factor after a division of two fixed-point numbers with the same scaling factor is 1. Again, the original scaling factor can be restored through bit shifting.

Apart from such fundamental operations, SW-UCB and D-UCB additionally require the square root function, as well as the natural logarithm function. The square root function can be efficiently realized based on two observations. First, we can translate the problem of computing the square root of a positive base-2 fixed-point number $a \times 2^x$ to the problem of computing the square root of the positive integer a. Since $\sqrt{a \times 2^x} = \sqrt{a} \times 2^{\frac{x}{2}}$, we only need to right-shift the binary result by $|\frac{x}{2}|$, which is unproblematic if x is even. Second, for computing $\sqrt{a}$, there is an efficient algorithm [1]. Let m be the greatest integer that fulfills $(2^m)^2 = 4^m \leq a$. The idea of this algorithm is to iterate over all bits $m, m - 1, \ldots, 0$ and set the i-th one in the intermediate result r if and only if $(r+2^i)^2 \leq a$. The natural logarithm function, on the other hand, can be restated using only binary logarithms. Let $a \times 2^x$ be a positive base-2 fixed-point number:

$$\log_e(a \times 2^x) = \frac{\log_2(a \times 2^x)}{\log_2(e)} = \frac{\log_2(a) + \log_2(2^x)}{\log_2(e)} \tag{5}$$

Thus, it suffices to have an efficiently implementable algorithm for computing $\log_2(a)$, which also exists [36]. That algorithm first computes the characteristic of $\log_2(a)$ as the number of divisions or multiplications by 2 that are necessary to normalize a into the interval $[1, 2)$. If divisions are necessary the characteristic is negative and else positive. Next, the mantissa bits are computed in a bit-by-bit manner, similar to the described square root algorithm.

4.2   Implementation Shortcuts

As for SW-UCB, the natural logarithm function is only invoked on Line 9 of Algorithm 2 with a numerical value from the set $\{K-1, K, \ldots, \tau\}$. Note that, in practice, τ can only be chosen rather small. This is because a node needs to store τ rewards along with the chosen channel for each neighbor. In the case of Bernoulli rewards, $N = 10$ neighbors, $K = 16$ channels, and $\tau = 100$ this already yields a RAM consumption of at least $(1 + \log_2(K)) \times \tau \times N = 625$ bytes. Hence, it is viable to use a lookup table for the natural logarithm function, which decreases the processing overhead of SW-UCB. On top of that, one multiplication can be saved by precomputing the multiplication with the exploration tendency ξ, too.

As for D-UCB, a naive implementation would store the whole history of rewards along with the selected channels in order to compute the discounted values on Line 5 and 6 of Algorithm 3. This can be avoided by storing the intermediary results and to multiply them by the discount factor γ after each time step. This implementation shortcut is listed as Algorithm 4. A potential pitfall is that, when using fixed-point arithmetic, n_k may become zero, which we suggest handling by pulling arm k, analogous to a similar issue in Algorithm 2.

Algorithm 4. D-UCB

Require: T: time horizon T
  K: number of arms
  $\xi > 0$: parameter for trading off exploitation against exploration
  $\gamma \in (0, 1)$: discount factor
1: $n_k \leftarrow 0$, for all $k \in [K]$
2: $x_k \leftarrow 0$, for all $k \in [K]$
3: **for** $t = 1$ to T **do**
4:    **if** $t \leq K$ **then**
5:      $I_t \leftarrow t$
6:    **else if** $n_k = 0$ for any $k \in [K]$ **then**
7:      $I_t \leftarrow k$
8:    **else**
9:      $I_t \leftarrow \arg\max_{k \in [K]} \frac{x_k}{n_k} + \sqrt{\frac{\xi \ln \sum_{l=1}^{K} n_l}{n_k}}$
10:   **end if**
11:   Pull arm I_t and receive reward $x_{I_t, t} \leftarrow X_{I_t, t}$
12:   $n_k \leftarrow n_k \times \gamma$, for all $k \in [K]$
13:   $x_k \leftarrow x_k \times \gamma$, for all $k \in [K]$
14:   $n_{I_t} \leftarrow n_{I_t} + 1$
15:   $x_{I_t} \leftarrow x_{I_t} + x_{I_t, t}$
16: **end for**

4.3   Integration into IEEE 802.15.4 CSL

CSL belongs to the class of asynchronous MAC protocols, which work without network-wide time synchronization. In the single-channel version of CSL, each node wakes up to scan for a so-called wake-up frame at regular intervals. In the multi-channel version of CSL, each node also wakes up at regular intervals. Yet, in contrast to single-channel CSL, a node not only scans for a wake-up frame on a single channel, but on all used channels, one after another. Thus, to keep the same base energy consumption as single-channel CSL, the wake-up interval has to be extended. A non-standard, yet beneficial approach to convert single-channel CSL to a multi-channel MAC protocol is to wake up on a different channel at every wake up, following a known channel hopping pattern [3,8]. This non-standard approach is beneficial in terms of latencies because it does not require extending the wake-up interval in order to keep the same base energy consumption as single-channel CSL. An implementation of non-standard multi-channel CSL is available for Contiki-NG and CC2538 SoCs [24]. We chose to integrate MAB-based channel hopping into that implementation of multi-channel CSL.

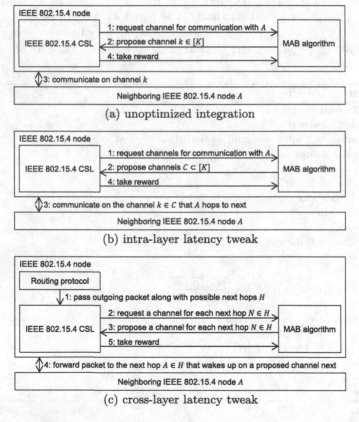

(a) unoptimized integration

(b) intra-layer latency tweak

(c) cross-layer latency tweak

Fig. 1. Integration of MAB-based channel hopping into CSL (a) without tweaks, (b) with an intra-layer latency tweak, and (c) with a cross-layer latency tweak

A simple integration of MAB-based channel hopping into non-standard multi-channel CSL is shown in Fig. 1a. When CSL has a frame to send to a particular neighbor A, the MAB algorithm is asked to select a channel $k \in [K]$ for that transmission. Subsequently, CSL waits until A is about to wake up on channel k and then sends the frame to A. Depending on if an acknowledgment frame is received or not, CSL reports back the reward 1 or 0, respectively.

Yet, waiting for a neighbor to wake up on a particular channel leads to increased latencies, at least when using non-standard multi-channel CSL. There are two ways to preserve the lower latencies to some extent. One is to ask the MAB algorithm to propose a subset of channels, enabling CSL to select the one that entails the lowest latency, as shown in Fig. 1b. Another requires cross-layer communication with the routing protocol. Specifically, in the case of a packet, the routing protocol may provide a subset of neighbors that makes progress toward the packet's destination. CSL can then pick the neighbor that will wake up next, as shown in Fig. 1c. Both remedies may be combined to achieve even lower latencies. For the scope of this paper, we restrict ourselves to realizing and comparing the options shown in Fig. 1a and 1b.

5 Evaluation

In this section, we first show that, in Monte Carlo simulations, SW-UCB achieves slightly higher PDRs than D-UCB, as well as that our implementations of SW-UCB and D-UCB perform equally well as floating-part counterparts. Then, we demonstrate that D-UCB attains slightly higher PDRs than SW-UCB in an intra-body communication use case. Finally, we shall find that D-UCB incurs a moderate RAM, program memory, and processing overhead on CC2538 SoCs, whereas SW-UCB's RAM consumption is impractical at adequate window sizes.

5.1 Monte Carlo Simulations

To investigate (i) possible performance discrepancies between our fixed-point and floating-point implementations of SW-UCB and D-UCB, as well as (ii) adequate parameter ranges, a simple Monte Carlo simulation environment was developed. It directly interfaced with our fixed-point and floating-point implementations of SW-UCB and D-UCB, similar as shown in Fig. 1a, except that the simulation environment replaced CSL. Throughout, the MAB algorithm was offered $K = 16$ channels to choose from and the time steps per simulation run were $T = 1000$. The probability of a successful frame transmission on a channel $k \in [K]$ followed a Bernoulli distribution with success probability p_k. The success probabilities randomly changed on all channels at change points and at the beginning of each simulation run. This randomization was done so that there was always one channel with a success probability 1 and the mean over all success probabilities was always $\frac{\sum_{k \in [K]} p_k}{K} = 0.5$. Three different environments were simulated, namely a stationary environment with no change points, a more dynamic environment with one change point at $t = 500$, and an even more dynamic environment with change points at time step $200, 400, 600$ and 800. As for SW-UCB, the simulated parameter settings

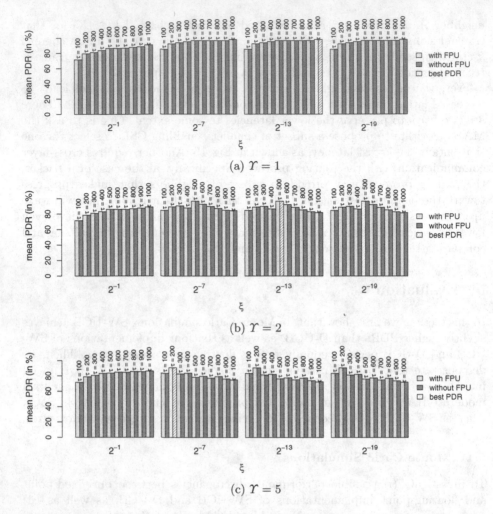

Fig. 2. Packet delivery ratios of SW-UCB in different simulated environments. In more dynamic environments, higher exploration tendencies ξ and lower window sizes τ lead to higher PDRs

were the exploration tendencies $\xi \in \{2^{-1}, 2^{-7}, 2^{-13}, 2^{-19}\}$ and the window sizes $\tau \in \{100, 200, \ldots, 1000\}$. As for D-UCB, the simulated parameter settings were the same exploration tendencies $\xi \in \{2^{-1}, 2^{-7}, 2^{-13}, 2^{-19}\}$ and the discount factors $\gamma \in \{1 - 2^{-2}, 1 - 2^{-3}, \ldots, 1 - 2^{-11}\}$. For each combination of parameters and environments 100,000 simulations were run.

Figure 2 shows the mean PDRs for SW-UCB. Throughout, the fixed-point implementation performs equally well as the floating-point implementation despite the floating-point implementation used 64-bit doubles, whereas the fixed-point implementation stored fixed-point numbers as 32-bit integers. This suggests that fixed-point arithmetic is suitable for implementing SW-UCB, provided

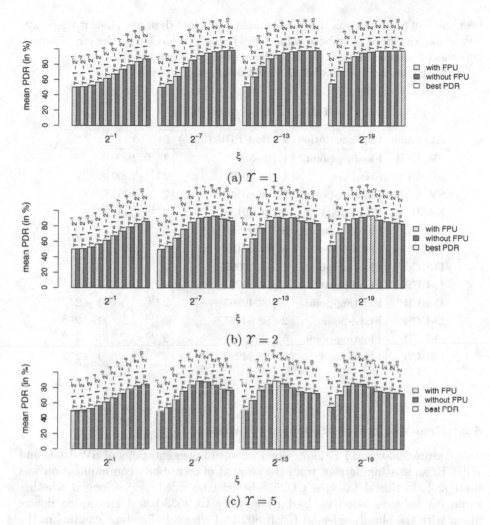

Fig. 3. Packet delivery ratios of D-UCB in different simulated environments. In more dynamic environments, higher exploration tendencies ξ and lower discount factors γ lead to higher PDRs

that numerically problematic ranges are avoided, which we did through additional scaling factors. Furthermore, the best setting of the window size τ matches the number of time steps between change points. This stresses that choosing τ as large as possible is suboptimal. The best value of the exploration tendency ξ increases as the environment gets more dynamic. This can be attributed to the fact that it pays off to explore other arms in a more dynamic environment.

Figure 3 shows the mean PDRs for D-UCB. Our fixed-point implementation of D-UCB also performed equally well as its floating-point counterpart, but we had to extend the length of mantissas from 16 bits to 22 bits. Unsurprisingly, the best value of ξ increases as the environment gets more dynamic since it then pays offs to explore more. The best value for γ tends to decrease as the environment gets more dynamic. This is because decreasing γ weighs recent awards more.

Table 1 gives an overview of the best parameter settings.

Table 1. Monte Carlo PDRs

Algorithm	Implementation	Υ	Best PDR (in %)	ξ	τ	γ
SW-UCB	Floating-point	1	98.50096	2^{-13}	1000	
SW-UCB	Fixed-point	1	98.50101	2^{-13}	1000	
SW-UCB	Floating-point	2	96.80545	2^{-19}	500	
SW-UCB	Fixed-point	2	96.80139	2^{-13}	500	
SW-UCB	Floating-point	5	90.99888	2^{-7}	200	
SW-UCB	Fixed-point	5	91.00135	2^{-7}	200	
D-UCB	Floating-point	1	98.50095	2^{-19}		$1 - 2^{-7}$
D-UCB	Fixed-point	1	98.50274	2^{-19}		$1 - 2^{-10}$
D-UCB	Floating-point	2	93.49205	2^{-19}		$1 - 2^{-6}$
D-UCB	Fixed-point	2	93.64187	2^{-19}		$1 - 2^{-6}$
D-UCB	Floating-point	5	88.93551	2^{-13}		$1 - 2^{-5}$
D-UCB	Fixed-point	5	88.94299	2^{-13}		$1 - 2^{-5}$

5.2   Real-World Packet Delivery Ratios

To measure real-world PDRs attained by our implementations of SW-UCB and D-UCB, an existing dataset from the context of intra-body communication was used [26]. In this dataset, a CC2538-based embedded device logged whether radio frames were acknowledged by another CC2538-based embedded device along with the blindly selected IEEE 802.15.4 channel. The best channel in the dataset has a mean PDR of 99.54% and the worst channel has a mean PDR of 71.19% (counting transmissions that were aborted due to a negative clear channel assessment as a failed transmission). This time, the experimental setup followed Fig. 1b, where CSL was replaced by the dataset in the following manner. When the MAB algorithm proposed channels $C \subset [16]$, the next sample in the dataset on a channel $k \in C$ was looked up and a reward of 1 was returned if that sample was successful and else 0. As soon as the dataset was exhausted, the simulation run ended. The same parameter settings like in the previous experiment were used. An additional parameter is the size of the subsets of channels $C \subset [16]$ that are being proposed by the MAB algorithm. It was set to $|C| \in \{1, 4\}$. For each combination of parameter settings, 100 samples were collected.

The results are shown in Fig. 4 and the best parameter settings are listed in Table 2. Accordingly, both SW-UCB and D-UCB achieve very high delivery ratios. However, the mean PDRs in the dataset are already 94.05% and

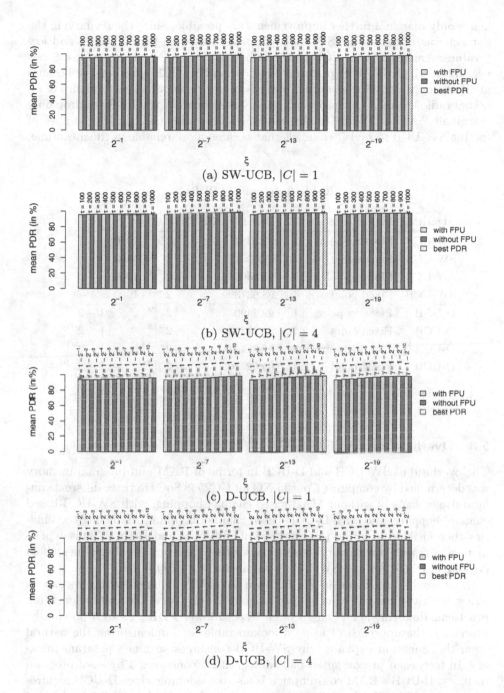

Fig. 4. Packet delivery ratios in the context of intra-body communication with and without the intra-layer latency tweak. The intra-layer latency tweak has only a small effect on PDRs while lowering latencies by a factor of $|C| = 4$.

hence only marginal further improvements are possible. Since the scenario in the dataset was rather static, SW-UCB performs best with high τ values and low ξ values. Analogously, D-UCB performs best with high γ values and low ξ values. Interestingly, D-UCB performs better than SW-UCB, presumably because D-UCB's approach of discounting rewards better reflects the performance of an actual radio channel. Our proposed latency tweak causes the PDRs to drop only marginally. As for SW-UCB, this tweak is even beneficial, maybe because of forcing SW-UCB to revisit channels that became more reliable in the meantime.

Table 2. Intra-body PDRs

Algorithm	Implementation	$|C|$	Best PDR (in %)	ξ	τ	γ
SW-UCB	Floating-point	1	98.5384	2^{-13}	1000	
SW-UCB	Fixed-point	1	98.6051	2^{-19}	1000	
SW-UCB	Floating-point	4	98.636894	2^{-13}	1000	
SW-UCB	Fixed-point	4	98.637399	2^{-13}	1000	
D-UCB	Floating-point	1	99.2499	2^{-19}		$1 - 2^{-9}$
D-UCB	Fixed-point	1	99.2145	2^{-13}		$1 - 2^{-10}$
D-UCB	Floating-point	4	98.972431	2^{-13}		$1 - 2^{-10}$
D-UCB	Fixed-point	4	99.027107	2^{-13}		$1 - 2^{-10}$

5.3   Overhead on CC2538 SoCs

The overhead of SW-UCB and D-UCB in terms of RAM and program memory was determined by compiling Contiki-NG for CC2538 SoCs in three different configurations, namely without MAB-based channel hopping, with SW-UCB-based channel hopping, and with D-UCB-based channel hopping. Analyzing the binaries then enabled isolating the overhead of SW-UCB and D-UCB. This experiment was repeated with varying settings for the maximum number of neighbors $N \in \{1, 6, 11, 16, 21\}$ and SW-UCB's parameter $\tau \in \{100, 500\}$.

The results are shown in Fig. 5. Since SW-UCB requires storing τ rewards along with the chosen channel per neighbor, only very low values for τ are practical. However, low τ values often lead to lower PDRs, as seen previously. Moreover, the approach of using a lookup table for implementing the natural logarithm function explains why SW-UCB consumes so much program memory. In fact, each precomputed natural logarithm consumes 4 bytes of program memory. D-UCB's RAM consumption is also considerable since D-UCB requires storing two 32-bit fixed-point numbers per each of the 16 channels and neighbor. Its program memory footprint, on the other hand, is quite low.

Lastly, to measure the processing overhead, two CC2538-based devices were programmed to send frames in a ping-pong manner. During this process, the

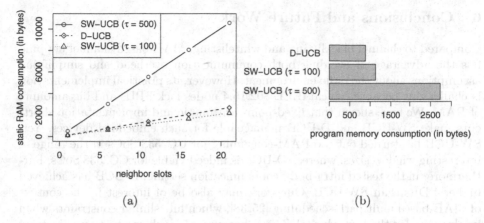

(a)   (b)

Fig. 5. Memory footprint of SW-UCB and D-UCB. Scaling τ up is impractical.

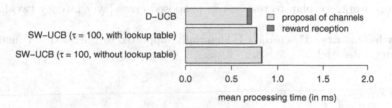

Fig. 6. Mean processing times of SW-UCB and D-UCB functions. D-UCB outperforms SW-UCB through the implementation shortcut of caching $\{n_i\}_{i \in [16]}$

sender side logged the processing time of proposing channels and of processing rewards. This experiment was conducted with both SW-UCB- and D-UCB-based channel hopping. Finally, in a third run of this experiment, SW-UCB's lookup table for natural logarithms was disabled. Throughout, τ was set to 100.

Figure 6 depicts the mean processing overhead. Surprisingly, even when SW-UCB uses precomputed natural logarithms, it still takes SW-UCB longer to propose channels compared to D-UCB. This can be attributed to the additional step of adding all rewards in the sliding window, whereas the equivalent part in D-UCB loads readily discounted rewards from RAM. On the other hand, this is why D-UCB needs slightly longer for processing rewards than SW-UCB.

SW-UCB's processing overhead is reducable by additionally precomputing:

$$\frac{x_k}{n_k} + \sqrt{\frac{\xi \ln \tau}{n_k}} \tag{6}$$

for all possible $n_k, x_k \in [\tau]$ and some fixed ξ. These precomputed values could then be used on Line 9 of Algorithm 2 at time steps $t - 1 \geq \tau$. Unfortunately, precomputing these values is only practical for a range of τ values that is uninteresting in terms of PDRs, at least as far as CC2538 SoCs are concerned.

6   Conclusions and Future Work

Compared to channel blacklisting and whitelisting, MAB-based channel hopping has the advantage of avoiding both communication overhead and simplifying assumptions about the radio environment. However, its practical implementation is challenging because typical IEEE 802.15.4 nodes lack FPUs and big amounts of RAM. We have shown that fixed-point arithmetic and implementation short-cuts make SW-UCB and D-UCB manageable for such embedded devices. Yet, SW-UCB has turned out too RAM-consuming for CC2538 SoCs in the range of interesting window sizes, whereas D-UCB is indeed viable for CC2538 SoCs. Furthermore, in the tested intra-body communication setting, D-UCB has achieved higher PDRs than SW-UCB. Our work may also be of interest in the context of MAB-based multipath scheduling [25,38], which hits similar constraints when implemented in the Linux kernel. This is because it is problematic to use floating-point instructions in kernel space and because kernel memory is not pageable. For future work, we plan to realize our proposed cross-layer latency tweak.

Acknowledgments. This project is financially supported by the Swedish Foundation for Strategic Research.

References

1. libfixmath. https://github.com/PetteriAimonen/libfixmath
2. IEEE Standard 802.15.4-2020 (2020)
3. Al Nahas, B., Duquennoy, S., Iyer, V., Voigt, T.: Low-power listening goes multi-channel. In: Proceedings of the 2014 IEEE International Conference on Distributed Computing in Sensor Systems (DCOSS), pp. 2–9. IEEE (2014). https://doi.org/10.1109/DCOSS.2014.33
4. Allesiardo, R., Féraud, R., Maillard, O.-A.: The non-stationary stochastic multi-armed bandit problem. Int. J. Data Sci. Anal. **3**(4), 267–283 (2017). https://doi.org/10.1007/s41060-017-0050-5
5. Auer, P., Cesa-Bianchi, N., Fischer, P.: Finite-time analysis of the multiarmed bandit problem. Mach. Learn. **47**(2–3), 235–256 (2002). https://doi.org/10.1023/A:1013689704352
6. Auer, P., Cesa-Bianchi, N., Freund, Y., Schapire, R.E.: The non-stochastic multi-armed bandit problem. SIAM J. Comput. **32**(1), 48–77 (2002). https://doi.org/10.1137/S0097539701398375
7. Besbes, O., Gur, Y., Zeevi, A.: Stochastic multi-armed-bandit problem with non-stationary rewards. In: Proceedings of the Neural Information Processing Systems Conference (NIPS 2014), pp. 199–207. Curran (2014)
8. Borms, J., Steenhaut, K., Lemmens, B.: Low-overhead dynamic multi-channel MAC for wireless sensor networks. In: Silva, J.S., Krishnamachari, B., Boavida, F. (eds.) EWSN 2010. LNCS, vol. 5970, pp. 81–96. Springer, Heidelberg (2010). https://doi.org/10.1007/978-3-642-11917-0_6
9. Cao, Y., Wen, Z., Kveton, B., Xie, Y.: Nearly optimal adaptive procedure with change detection for piecewise-stationary bandit. In: Proceedings of the 22nd International Conference on Artificial Intelligence and Statistics (AISTATS), pp. 418–427. PMLR (2019)

10. Cheung, W.C., Simchi-Levi, D., Zhu, R.: Learning to optimize under non-stationarity. In: Proceedings of the 22nd International Conference on Artificial Intelligence and Statistics (AISTATS), pp. 1079–1087. PMLR (2019)
11. Dakdouk, H., Tarazona, E., Alami, R., Féraud, R., Papadopoulos, G.Z., Maillé, P.: Reinforcement learning techniques for optimized channel hopping in IEEE 802.15.4-TSCH networks. In: Proceedings of the 21st ACM International Conference on Modeling, Analysis and Simulation of Wireless and Mobile Systems (MSWIM 2018), pp. 99–107. ACM (2018). https://doi.org/10.1145/3242102.3242110
12. Du, P., Roussos, G.: Adaptive time slotted channel hopping for wireless sensor networks. In: Proceedings of the 2012 4th Computer Science and Electronic Engineering Conference (CEEC 2012), pp. 29–34. IEEE (2012). https://doi.org/10.1109/CEEC.2012.6375374
13. Elsts, A.: Source-node selection to increase the reliability of sensor networks for building automation. In: Proceedings of the 2016 International Conference on Embedded Wireless Systems and Networks (EWSN 2016), pp. 125–136 (2016)
14. Elsts, A., Fafoutis, X., Piechocki, R., Craddock, I.: Adaptive channel selection in IEEE 802.15.4 TSCH networks. In: Proceedings of the 2017 Global Internet of Things Summit (GIoTS), pp. 1–6. IEEE (2017). https://doi.org/10.1109/GIOTS.2017.8016246
15. Farahmandand, M., Nabi, M.: Channel quality prediction for TSCH blacklisting in highly dynamic networks: a self-supervised deep learning approach. IEEE Sens. J. 21(18), 21059–21068 (2021). https://doi.org/10.1109/JSEN.2021.3093424
16. Garivier, A., Moulines, E.: On upper-confidence bound policies for switching bandit problems. In: Kivinen, J., Szepesvári, C., Ukkonen, E., Zeugmann, T. (eds.) ALT 2011. LNCS (LNAI), vol. 6925, pp. 174–188. Springer, Heidelberg (2011). https://doi.org/10.1007/978-3-642-24412-4_16
17. Gomes, P.H., Watteyne, T., Krishnamachari, B.: MABO-TSCH: multihop and blacklist-based optimized time synchronized channel hopping. Trans. Emerg. Telecommun. Technol. 29(7) (2018). https://doi.org/10.1002/ett.3223
18. Gürsu, M., Vilgelm, M., Zoppi, S., Kellerer, W.: Reliable co-existence of 802.15.4e TSCH-based WSN and Wi-Fi in an aircraft cabin. In: Proceedings of the 2016 IEEE International Conference on Communications Workshops (ICC 2016), pp. 663–668. IEEE (2016). https://doi.org/10.1109/ICCW.2016.7503863
19. Hartland, C., Gelly, S., Baskiotis, N., Teytaud, O., Sebag, M.: Multi-armed Bandit, Dynamic Environments and Meta-Bandits (2006)
20. Incel, O.D., Dulman, S., Jansen, P.: Multi-channel support for dense wireless sensor networking. In: Havinga, P., Lijding, M., Meratnia, N., Wegdam, M. (eds.) EuroSSC 2006. LNCS, vol. 4272, pp. 1–14. Springer, Heidelberg (2006). https://doi.org/10.1007/11907503_1
21. Javan, N.T., Sabaei, M., Hakami, V.: Adaptive channel hopping for IEEE 802.15.4 TSCH-based networks: a dynamic Bernoulli Bandit approach. IEEE Sens. J., 1 (2021). https://doi.org/10.1109/JSEN.2021.3110720
22. Kim, Y., Shin, H., Cha, H.: Y-MAC: an energy-efficient multi-channel MAC protocol for dense wireless sensor networks. In: Proceedings of the 2008 International Conference on Information Processing in Sensor Networks (IPSN 2008), pp. 53–63. IEEE (2008). https://doi.org/10.1109/IPSN.2008.27
23. Kocsis, L., Szepesvári, C.: Discounted UCB. In: Proceedings of the 2nd PASCAL Challenges Workshop (2006)

24. Krentz, K.-F.: A denial-of-sleep-resilient medium access control layer for IEEE 802.15.4 networks. Ph.D. thesis, Potsdam University (2019). https://doi.org/10.25932/publishup-43930

25. Krentz, K.-F., Corici, M.-I.: Poster: multipath extensions for WireGuard. In: Proceedings of the 2021 IFIP Networking Conference (IFIP Networking), pp. 1–3. IEEE (2021). https://doi.org/10.23919/IFIPNetworking52078.2021.9472775

26. Krentz, K.-F., Padmal, M., Mandal, B., Augustine, R., Voigt, T.: Dataset: enabling offline tuning of fat channel communication. In: Proceedings of the Data: Acquisition to Analysis (DATA 2021). ACM (2021). https://doi.org/10.1145/3485730.3494112

27. Li, P., Vermeulen, T., Liy, H., Pollin, S.: An adaptive channel selection scheme for reliable TSCH-based communication. In: Proceedings of the 2015 International Symposium on Wireless Communication Systems (ISWCS 2015), pp. 511–515. IEEE (2015). https://doi.org/10.1109/ISWCS.2015.7454397

28. Liu, F., Lee, J., Shroff, N.B.: A change-detection based framework for piecewise-stationary multi-armed bandit problem. CoRR abs/1711.03539 (2017)

29. Luo, J., Su, X., Liu, B.: A reinforcement learning approach for multipath TCP data scheduling. In: Proceedings of the 2019 IEEE 9th Annual Computing and Communication Workshop and Conference (CCWC), pp. 276–280. IEEE (2019). https://doi.org/10.1109/CCWC.2019.8666496

30. Mellor, J., Shapiro, J.: Thompson sampling in switching environments with Bayesian online change detection. In: Proceedings of the 16th International Conference on Artificial Intelligence and Statistics (AISTATS), pp. 442–450. PMLR (2013)

31. Oksanen, J., Koivunen, V.: Learning spectrum opportunities in non-stationary radio environments. In: Proceedings of the 2017 IEEE International Conference on Acoustics, Speech and Signal Processing (ICASSP), pp. 2447–2451. IEEE (2017). https://doi.org/10.1109/ICASSP.2017.7952596

32. Pister, K., Doherty, L.: TSMP: time synchronized mesh protocol. IASTED Distrib. Sens. Netw. **391**(398), 61 (2008)

33. Salajegheh, M., Soroush, H., Kalis, A.: HYMAC: hybrid TDMA/FDMA medium access control protocol for wireless sensor networks. In: Proceedings of the 2007 IEEE 18th International Symposium on Personal, Indoor and Mobile Radio Communications (PIMRC 2007), pp. 1–5. IEEE (2007). https://doi.org/10.1109/PIMRC.2007.4394374

34. Tavakoli, R., Nabi, M., Basten, T., Goossens, K.: Enhanced time-slotted channel hopping in WSNs using non-intrusive channel-quality estimation. In: Proceedings of the 2015 IEEE 12th International Conference on Mobile Ad Hoc and Sensor Systems (MASS 2015), pp. 217–225. IEEE (2015). https://doi.org/10.1109/MASS.2015.48

35. Texas Instruments: CC2538 SoC for 2.4-GHz IEEE 802.15.4 & ZigBee/ZigBee IP Applications User's Guide (Rev. C). http://www.ti.com/lit/ug/swru319c/swru319c.pdf

36. Turner, C.S.: A fast binary logarithm algorithm. IEEE Signal Process. Mag. **27**(5), 124–140 (2010). https://doi.org/10.1109/MSP.2010.937503

37. Wood, A., Stankovic, J., Zhou, G.: DEEJAM: defeating energy-efficient jamming in IEEE 802.15.4-based wireless networks. In: Proceedings of the 4th Annual IEEE Communications Society Conference on Sensor, Mesh and Ad Hoc Communications and Networks (SECON 2007), pp. 60–69. IEEE (2007). https://doi.org/10.1109/SAHCN.2007.4292818

38. Wu, H., Alay, Ö., Brunstrom, A., Ferlin, S., Caso, G.: Peekaboo: learning-based multipath scheduling for dynamic heterogeneous environments. IEEE J. Sel. Areas Commun. **38**(10), 2295–2310 (2020). https://doi.org/10.1109/JSAC.2020.3000365
39. Yu, J.Y., Mannor, S.: Piecewise-stationary bandit problems with side observations. In: Proceedings of the 26th Annual International Conference on Machine Learning (ICML 2009), pp. 1177–1184. ACM (2009). https://doi.org/10.1145/1553374.1553524
40. Zhou, L.: A survey on contextual multi-armed bandits. arXiv preprint arXiv:1508.03326 (2015)
41. Zorbas, D., Papadopoulos, G.Z., Douligeris, C.: Local or global radio channel blacklisting for IEEE 802.15.4-TSCH networks? In: Proceedings of the 2018 IEEE International Conference on Communications (ICC 2018), pp. 1–6. IEEE (2018). https://doi.org/10.1109/ICC.2018.8423007

Cross Inference of Throughput Profiles Using Micro Kernel Network Method

Nageswara S. V. Rao[1(✉)], Anees Al-Najjar[1], Neena Imam[1], Zhengchun Liu[2], Rajkumar Kettimuthu[2], and Ian Foster[2]

[1] Oak Ridge National Laboratory, Oak Ridge, TN, USA
{raons,alnajjarm,nimam}@ornl.gov
[2] Argonne National Laboratory, Argonne, IL, USA
{zhengchun.liu,kettimut,foster}@anl.gov

Abstract. Dedicated network connections are being increasingly deployed in cloud, centralized and edge computing and data infrastructures, whose throughput profiles are critical indicators of the underlying data transfer performance. Due to the cost and disruptions to physical infrastructures, network emulators, such as Mininet, are often used to generate measurements needed to estimate throughput profiles, typically expressed as a function of the connection round trip time. The profiles estimated using measurements from such emulated networks are usually inaccurate for high bandwidth and high latency connections, since they do not accurately reflect the critical network transport dynamics mainly due to computing and memory constraints of the host. We present a machine learning (ML) method to estimate the throughput profiles using emulation measurements to closely match the testbed and production network profiles. In particular, we propose a micro Kernel Network (mKN) that provides baseline throughput measurements on the host running Mininet emulations, which are used to learn a regression map that converts them to the corresponding testbed measurement estimates. Once initially learned, this map is applied to measurements from subsequent network emulations on the same host. We present experimental measurements to illustrate this approach, and derive generalization equations for the proposed mKN-ML method. Using a four-site scenario emulation, we show the effectiveness of this method in providing accurate concave throughput profiles from inaccurate convex or non-smooth ones indicated by Mininet emulation.

This work is performed at Oak Ridge National Laboratory managed by UT-Battelle, LLC for U.S. Department of Energy under Contract No. DE-AC05-00OR22725, and Argonne National Laboratory under Contract No. DE-AC02-06CH11357. The United States Government retains and the publisher, by accepting the article for publication, acknowledges that the United States Government retains a nonexclusive, paid-up, irrevocable, world-wide license to publish or reproduce the published form of this manuscript, or allow others to do so, for United States Government purposes. The Department of Energy will provide public access to these results of federally sponsored research in accordance with the DOE Public Access Plan (http://energy.gov/downloads/doe-public-access-plan).

© Springer Nature Switzerland AG 2022
É. Renault et al. (Eds.): MLN 2021, LNCS 13175, pp. 48–68, 2022.
https://doi.org/10.1007/978-3-030-98978-1_4

Keywords: Data transport infrastructure · Throughput profile · Network emulation · Round trip time · Machine learning · Generalization bounds · Regression estimates

1   Introduction

High performance data transport infrastructures with Data Transfer Nodes (DTNs) are increasingly being deployed to support cloud, centralized and edge computing applications in scientific and commercial scenarios, for example, Google data centers worldwide [11] and national laboratories over ESnet [1]. Dedicated connections between DTNs provide unimpeded capacity that can be exploited by the underlying data and file transfer software, which in turn requires optimal configurations of Transmission Control Protocol (TCP). The TCP throughput profile of an infrastructure, typically expressed as a function of round trip time (RTT), captures its performance in terms of both network and host configurations; in particular, its shape provides critical performance information. The peak throughput is a direct indicator of performance, and the concave-convex geometry is a more subtle indicator [13], with a concave (convex) profile indicating intermediate-RTT throughput higher (lower) than linear interpolations. For an infrastructure, a smooth and concave profile similar to Fig. 1(a) is desired, which is achieved by optimizing the hardware and software of all DTNs often under identical configurations.

Measurements collected over physical and emulated connections are used to generate profiles to assess various existing and future envisioned infrastructures, respectively. In particular, software and testbed emulations have been useful in assessing the designs for future data transport networks that underlie general networked infrastructures and federations. Measurements on physical infrastructures are expensive and time-consuming to collect, and often are intrusive and consume resources; furthermore, they are not available while these infrastructures are being designed and built. Both hardware and software emulators are used to estimate approximations to the profiles of physical, production infrastructures, and they additionally support design exploration and optimizations. In particular, Mininet has been extensively used for emulating complex networks, and has been an invaluable tool for developing various softwarization tools and frameworks, for example, the ContainerNet. Also, testbeds with dedicated hosts (identical to production DTNs) connected over hardware connection emulators have been useful in estimating profiles for a wide range of RTTs, including those not readily available in physical production and testbed networks.

1.1   Profiles from Network Emulations

The Mininet emulations in software, however, are limited in providing performance profiles of high performance transport networks [15]; for example, basic link classes (e.g., TClink) are limited to 1 Gbps capacity and host-based 10 Gbps and higher extensions based on Token Bucket Filters (TBF) [3] do not closely match the

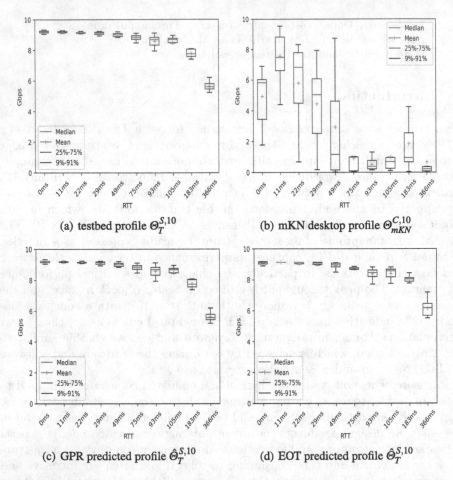

(a) testbed profile $\Theta_T^{S,10}$

(b) mKN desktop profile $\Theta_{mKN}^{C,10}$

(c) GPR predicted profile $\hat{\Theta}_T^{S,10}$

(d) EOT predicted profile $\hat{\Theta}_T^{S,10}$

Fig. 1. Throughput profiles of testbed and Mininet emulation on desktop workstation, and GPR and EOT estimates

physical measurements. While they are accurate for smaller messages (e.g., control messages, small data transfers), they are not very reliable when host resources are insufficient to sustain the network flows associated with high-bandwidth wide-area transfers. Indeed, their profiles can be misleading, as shown in Fig. 1(b) for a network configuration identical to that used in Fig. 1(a). Indeed, the latter is mostly convex which indicates performance bottlenecks, and non-smooth which indicates under performance by some hosts [12]; but, these hosts are identically configured with the same parameters as in the former. At the core, this phenomenon is a result of TCP dynamics not being replicated well in Mininet emulations, which are typically run in a virtual machine (VM) on a host. This is a result of host system being shared by all concurrent TCP flows of the entire emulated network, in contrast with a DTN that typically supports a single server or client. Indeed, the effects of host CPU and memory limits become more dominant for large emulated networks, and they vary drastically between laptops and high-end servers when used as hosts. Yet, these TCP dynamics are critical in determining the precise concave-convex geometry of throughput profiles.

Hardware testbed emulators that employ dedicated DTN hosts connected over hardware connection emulators provide measurements that closely match the throughput of physical infrastructures, for example, within 2% for 10 Gbps wide-area connections [16]. They are typically limited to single connections and are not practical in emulating large, complex networks that Mininet is suited for. These *testbed profiles*, nevertheless, adequately capture the performance of physical infrastructures since hosts and connections closely match their physical counterparts, and hence are used as target profiles in this paper.

1.2 Contributions

We present a cross inference method to estimate testbed profiles from Mininet measurements using Machine Learning (ML) methods. We propose a *micro Kernel Network* (mKN) which implements a basic network to provide baseline measurements specific to a host. These measurements are used together with testbed measurements to learn a map that converts the former to latter, thereby providing profiles that closely match physical infrastructure profiles. This process is carried out in an initial step on the host by collecting mKN measurements for a suitable range of RTTs used in emulations. Subsequently, the learned map is applied to measurements from emulations running on the same host. The estimated profile by the mKN-ML method closely matches the testbed profile, for example, as indicated by profiles in Fig. 1(c) and Fig. 1(d) obtained using the Gaussian Process Regression (GPR) and Ensemble of Trees (EOT) ML regressions, respectively.

We present experimental measurements to illustrate the mKN-ML method, and derive its distribution-free generalization equations [19]. Using a four-site emulation scenario, we show the effectiveness of this method in providing an accurate concave profile from an inaccurate convex profile obtained from its Mininet emulation.

The organization of this paper is as follows. We briefly discuss some basic properties of TCP throughput profiles in Sect. 2. The mKN concept and its Mininet implementation, and the mKN-ML cross inference method are described in Sect. 3. Experimental scenarios are described in Sect. 4, and the estimated profiles are described in Sect. 5. The generalization equations and conclusions are presented in Sects. 6 and 7, respectively.

2 TCP Throughput Profile

The TCP throughput profile, which is expressed as a function of RTT, is determined by:

(i) host properties including protocol parameters including its version $V = C$, B, H, HS, I, S, and W, representing CUBIC [14], BBR [6], Hamilton TCP

(HTCP) [17], HighSpeed TCP [8], Illinois, Scalable TCP [10] and Westwood TCP versions, respectively, the number of parallel streams n, and various buffer sizes considered to be set for high throughput, and

(ii) connection properties including RTT, modality, and capacity, e.g., 10 Gbps for 10 GigE and 9.6 Gbps for SONET OC192.

Let $\theta_M^{V,n}(\tau, t)$ denote the aggregate throughput at time t over a connection of RTT τ, where M corresponds to origin of measurements, namely, mKN and T for testbed. These parameters are explicitly used to denote the throughput profile

$$\Theta_M^{V,n}(\tau) = \frac{1}{T_O} \int_0^{T_O} \theta_M^{V,n}(\tau, t) dt,$$

and selectively suppressed when evident from the context. The throughput dynamics of a transport with fixed parameters over a connection with RTT τ and capacity C are characterized by two phases. In the ramp-up phase, $\theta(t)$ increases for a duration T_R until it reaches a peak and then switches to a sustained throughput phase, wherein throughput is "sustained" using a mechanism which processes the acknowledgments, and infers and responds to losses. The average throughput is

$$\Theta_M(\tau) = \frac{T_R}{T_O}\bar{\theta}_R(\tau) + \frac{T_S}{T_O}\bar{\theta}_S(\tau) = \bar{\theta}_S(\tau) - f_R\left(\bar{\theta}_S(\tau) - \bar{\theta}_R(\tau)\right),$$

where $T_O = T_R + T_S$ and $f_R = T_R/T_O$.

A function $f(\tau)$ is *concave* [5] in interval I if for any $\tau_1 < \tau_2 \in I$, the following condition is satisfied: for $x \in [0, 1]$

$$f\left(x\tau_1 + (1-x)\tau_2\right) \geq xf(\tau_1) + (1-x)f(\tau_2).$$

It is *convex* if $\geq$ in the above condition is replaced by $\leq$. A function is concave if and only if $\frac{df}{d\tau}$ is a non-increasing function of τ or equivalently $\frac{d^2f}{d\tau^2} \leq 0$. For data transport infrastructures, typically, the throughput profile is concave when RTT is small, and at transition RTT τ_T it becomes and continues to be convex as RTT is increased [13].

For highly optimized infrastructures, the profile is entirely concave as shown in Fig. 1(a) for testbed connections. At the other extreme, an entirely convex profile indicates a performance bottleneck such as insufficient buffer size or processor cycles, as shown in Fig. 7(b); here, the processing capability of host limits the throughput in Mininet emulations. Typically, non-smooth profiles indicate that a subset (but not all) DTNs are under-optimized in physical infrastructures [12]. In particular, when all DTNs are identically optimized, their profiles align, thereby resulting in a smooth infrastructure profile; it will be concave if all are fully optimized and convex if all are identically under-optimized. When some DTNs are under-optimized, the throughput corresponding to their connection RTTs are lower compared to those of fully-optimized DTNs, thereby resulting in dips in throughput and an overall non-smooth profile. Thus, the mKN profile in Fig. 1(b) is misleading on two counts, namely, the overall convexity and non-smoothness, since its virtual hosts (vhosts) in Mininet use an identical DTN configuration of testbed hosts set for high throughput.

3   mKN-ML Method

The mKN is a basic network with programmable RTT shown in Fig. 2 that enables throughput measurements to be collected for a collection of RTTs of LAN and WAN connections. It is implemented in Mininet wherein vhosts are configured as DTNs; since vhosts inherit OS environment of the host they represent identical DTNs with tuned TCP parameters. Its baseline measurements are collected once at the beginning on the host, and subsequently used in the cross inference of measurements from emulations of possibly several different networks on the same host.

The cross inference ML method employs a function $\hat{f}$ to map mKN baseline measurements to testbed measurements as shown in Fig. 3. The regression map $\hat{f}$ is learned using mKN measurements that have been collected for a set of RTTs chosen to represent the emulated networks on the host, and the measurements collected over the testbed at same RTTs. The model can also be learned using historic measurements that have been collected over testbeds and production networks for various configurations, including different TCP versions and connection modalities. We apply two ML methods to train the map $\hat{f}$, namely, using smooth GPR and non-smooth EOT methods that represent two different approaches. This map is estimated once at the beginning using mKN measurements, and applied to subsequent measurements from emulated networks on the same host.

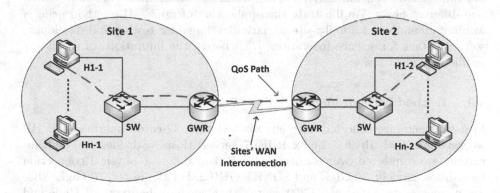

Fig. 2. mKN and its Mininet implementation.

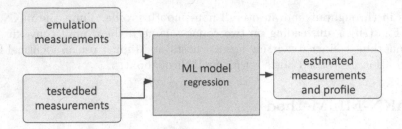

Fig. 3. Cross inference mKN-ML method outline.

We consider a generic set of RTTs, $\tau \in \{1, 11, 22, 29, 49, 75, 93, 105, 150, 182, 366\}$ ms, for which several data sets exist, and also collect additional data using our testbed with hardware connection emulators. The lower RTTs represent the cross-country connections, as between facilities across the US; the 93 and 182 ms RTTs represent inter-continental connections, as between US, Europe, and Asia; and the 366 ms RTT represents a hypothetical connection spanning the globe. We execute these Mininet emulations in a VirtualBox VM with Ubuntu LTS 18.4 OS under Windows10 workstation and laptop host systems. We use the GPR and EOT codes from scikitlearn Python libraries. We also tested GPR and EOT implementations from matlab statistics toolbox with very similar (almost identical) results.

4   Testbed and Emulation Measurements

The testbed measurements used to estimate the mKN-ML regression $\hat{f}$ are collected over connections consisting of dedicated hosts and connection realized by hardware emulators. The corresponding Mininet measurements are collected on two different hosts. We illustrate the application of mKN-ML method using a Mininet emulation of a multi-site scenario developed for testing and developing software stack for science federations [7], wherein the limitations of emulation profiles are identified.

4.1   Testbed Measurements

Measurements are collected over our testbed with 32-core Supermicro or HP workstations with Redhat Linux RHEL7 kernel. Hosts with identical configurations are connected over connections with two different physical connection modalities, namely, 10 GigE and SONET OC192. For the latter, 10 GigE NICs are connected to a Force10 E300 switch that converts between 10 GigE and SONET frames, and the OC192 ANUE emulator is in turn connected to the WAN ports of the E300, as shown in the top connection in Fig. 4(a). We use suites of emulated 10 GigE and SONET/OC192 connections via ANUE devices for the generic RTT set, and each measurement is repeated 10 times. We collected TCP memory-to-memory throughput measurements for multiple TCP

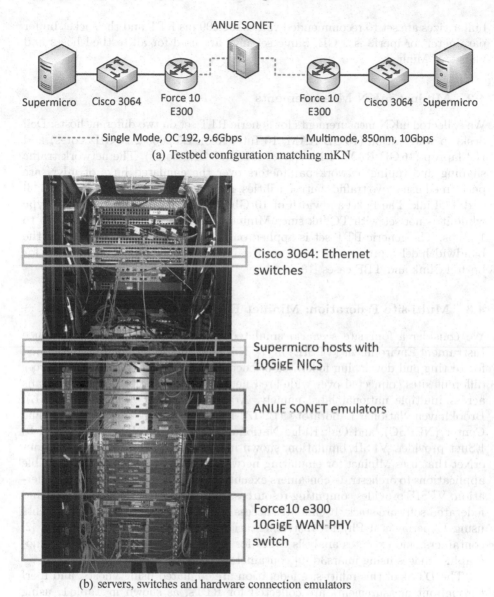

(a) Testbed configuration matching mKN

(b) servers, switches and hardware connection emulators

Fig. 4. Physical testbed for target measurements.

congestion control modules using the *iperf3*. We consider CUBIC, BBR, Hamilton TCP, Illinois, Scalable TCP and Westwood TCP versions in 15 scenarios shown in Table 2. The number of parallel streams is varied from 1 to 10 for each configuration, and throughput measurements are repeated 10 times. The collection of measurements is automated using bash scripts that use curl to configure ANUE emulators; for the generic RTT set, it typically takes 1–2 days. TCP buffer sizes are set to recommended values for 200 ms RTT and the socket buffer parameter for iperf3 is 2 GB. Same settings are used for all testbed hosts and vhosts of Mininet.

4.2   Mininet mKN Measurements

We collected mKN measurements for generic RTT set on two different hosts: Dell desktop workstation (16 GB RAM, 10 Intel Xeon E5-2630 2.20 GHz cores) and HP laptop (16 GB RAM, 4 Intel i7-8565U 1.80 GHz cores). The network traffic shaping and tuning network parameters over the emulated links of mKN are performed using two traffic control utilities, namely Token Buffer Filter (TBF) [3] and TCLink. The link bandwidth of 10 Gbps is set with TBF connection type while it is not set with TClink since Mininet limits its maximum bandwidth to 1 Gbps. The generic RTT set is applied on the connections while adapting the bandwidth delay product for TCP autotuning based on RTT values on hosts for both TClink and TBF cases [18].

4.3   Multi-site Federation: Mininet Emulation

We consider a four site scenario emulated by the Virtual Federated Science Instrument Environment (VFSIE) in [7]. It is an emulation framework developed for testing and developing federated science workflows that are executed across different sites connected over wide-area networks. This notional scenario spans across multiple national labs, namely, Argonne National Laboratory (ANL), Brookhaven National Laboratory (BNL), National Energy Research Scientific Center (NERSC), and Oak Ridge National Laboratory (ORNL) connected to ESnet provider. VFSIE emulation, shown in Fig. 5, is developed using ContainerNet that uses Mininet for emulating networks and hosts, Docker, and Ansible applications to orchestrate containers execution on virtual hosts across the federation. VFSIE provides computing resources to execute the science workflows and federated software stack. It emulates access for applications to steer instruments using Experimental Physics Instrumentation and Control system (EPICS) [2] containers, and executes analysis codes for accumulated neutron/photon tomographic images using imars3d [9] containers.

The DTNs of this multi-site federation are emulated using vhosts, and iperf throughput measurements are collected for RTTs, as shown in Table 1, using Ansible framework. There RTT values represent the notional values that nearly match but not necessarily identical to those of the corresponding physical connections. These profiles deviated significantly from those estimated from physical and testbed measurements with corresponding RTTs, and furthermore they

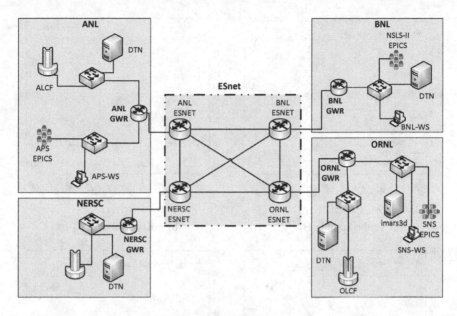

Fig. 5. Mininet emulation of four-site federation in VFSIE [7].

Table 1. Notional RTT between sites.

Connection	RTT (ms)
ANL–ORNL	22
ANL–BNL	75
BNL–ORNL	93
ANL–NERSC	105
ORNL–NERSC	150
BNL–NERSC	183

varied when VFSIE VM is executed on different hosts. On a single host, these profiles are generally preserved, except for some statistical variations, when Mininet is executed over connections of same RTT. We aim to estimate the throughput profile of DTNs at four sites using measurements collected over VFSIE. As will be shown in next section the mKN-ML map $\hat{f}$ will be used to convert these VFSIE measurements, thereby generating profiles that match testbed profiles.

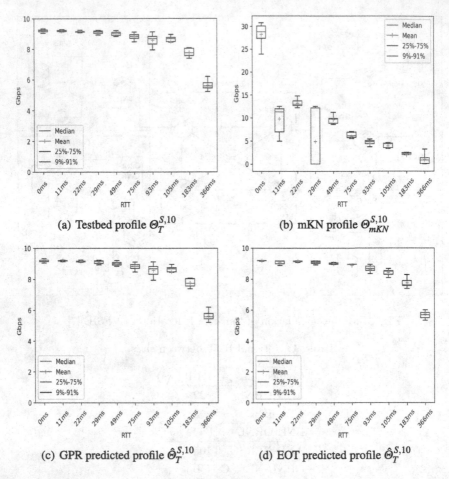

(a) Testbed profile $\Theta_T^{S,10}$ 　　　　　　 (b) mKN profile $\Theta_{mKN}^{S,10}$

(c) GPR predicted profile $\hat{\Theta}_T^{S,10}$ 　　　　 (d) EOT predicted profile $\hat{\Theta}_T^{S,10}$

Fig. 6. Testbed (S), mKN (S) and estimated target (S) profiles for generic RTT set (Table 2: row 7).

5   Estimated Profiles

Mininet measurement are collected for mKN and VFSIE connections, and the testbed measurements are collected for the corresponding RTTs. In each case, we collected iperf measurements for 1–10 parallel TCP streams, and each measurement is repeated 10 times. These measurements are used by the mKN-ML method to learn the map $\hat{f}$ using both GPR and EOT methods. There methods are chosen for their diversity of designs, namely, smoothness of GPR and non-smoothness of EOT regression; their Root Mean Square Error (RMSE) is comparable despite the difference. A summary of 15 illustrative cases with 10 parallel flows is presented in Table 2 for the target profiles with six (C, I, H, HS, S and W) TCP versions over OC192 testbed connections; the results for

Table 2. Summary: mKN and estimated profiles.

No	Host:Traffic Control Type	mKN: Θ_{mKN}	$\hat{f}$	EOT	GPR
1	Desktop:TClink	NS	C→C	0.2759	0.1189
2	Desktop:TClink	NS	H→ HS^a	0.1145	0.0649
3	Desktop:TClink	NS	H→ HS	0.6013	0.4270
4	Desktop:TBF	NS	S→S	0.2279	0.1485
5	Desktop:TBF	NS	S→W	0.8436	0.7026
6	Desktop:TBF	NS	S→I	0.8562	0.7198
7	Laptop:TClink	CVX, NS	S→S	0.1539	0.0176
8	Laptop:TClink	CVX	B→ C	0.9878	0.7460
9	Laptop:TBF	CVX	B→ C	0.6503	0.3940
10	Laptop:TBF	NS	C→S	0.2915	0.0118
11	Laptop:TBF	NS	S→S	0.2555	0.0946
12	VFSIE:TClink	CVX	C→C	0.0999	0.0279
13	VFSIE:TClink	CVX	C→S	0.0728	0.0228
14	VFSIE:TClink	CVX	C→W	0.0046	0.0013
15	VFSIE:TClink	CVX	C→I	0.0060	0.0018

C: CUBIC, S: STCP, W: Westwood, I: Illinois, B:BBR, H: HTCP, HS
:HighSpeed; MTU: 9K
^{a}: default TCP parameters
CVX: CONVEX, NS: NON SMOOTH
EOT and GPR columns provide RMSE

other TCP versions and 10 GigE connections are qualitatively similar and not included here due to page limit.

5.1   mKN Generic RTT Set: Concave Target Profiles

The mKN profiles showed significant variations, but mainly non-smooth (NS) or convex (CVX) or both, among the combinations as shown in Table 2. These profiles, however, are stable when repeated for a fixed configuration on a host. These for the generic RTT set for HP laptop are shown in Fig. 1(b) and Fig. 6(b) for CUBIC and STCP, respectively. The target profiles based on testbed measurements are concave for both CUBIC and STCP: see Fig. 1 (a), Fig. 6(a), and Fig. 7(a) for the latter. The estimated testbed profiles by applying ML map to these measurements are shown in Fig. 1(c) and Fig. 6(c) using EOT method, and in Fig. 1(d) and Fig. 6(d) using GPR method; the concave target profile is predicted in both cases by both ML methods. In all cases, the estimated profiles

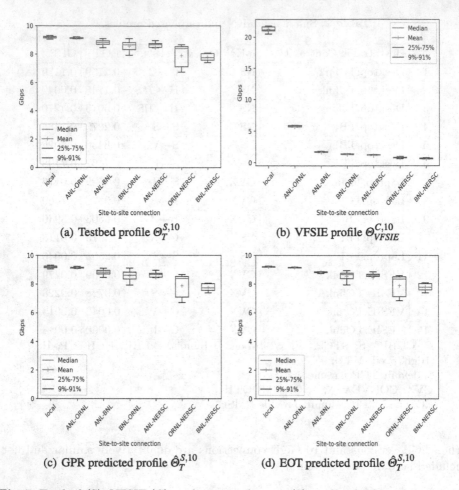

(a) Testbed profile $\Theta_T^{S,10}$

(b) VFSIE profile $\Theta_{VFSIE}^{C,10}$

(c) GPR predicted profile $\hat{\Theta}_T^{S,10}$

(d) EOT predicted profile $\hat{\Theta}_T^{S,10}$

Fig. 7. Testbed (S), VFSIE (C), and estimated target (S) profiles for four site scenario (Table 2: row 13).

are a close approximation to the testbed profiles with RMSE shown in Table 2 for Python libraries, and similar results are obtained for Matlab toolkit[1].

5.2   VFSIE Measurements: Concave Target Profiles

We collected iperf measurements between the DTNs at sites whose notional connection RTTs are shown in Table 1, using Mininet default link parameters. The VFSIE profile shown in Fig. 7(b) is smooth but convex, which is misleading since it does not represent the expected concave profile. Indeed, this convex

---

[1] While both libraries utilize the same basic methods, the EOT and GPR codes are sometimes found to yield varying performances in RMSE and eventual convergence, and these data sets in our case are robust to such effects.

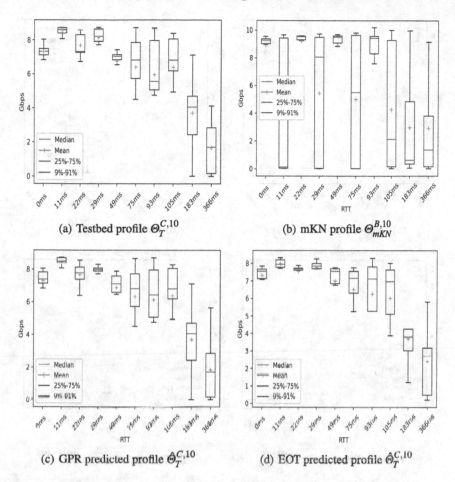

(a) Testbed profile $\Theta_T^{C,10}$ (b) mKN profile $\Theta_{mKN}^{B,10}$

(c) GPR predicted profile $\hat{\Theta}_T^{C,10}$ (d) EOT predicted profile $\hat{\Theta}_T^{C,10}$

Fig. 8. mKN on laptop with TBF and BBR, and testbed with CUBIC (Table 2: row 9).

profile is indicative of performance bottleneck caused by TCP threads not being serviced by host processes. The mapping of CUBIC to STCP measurements generates accurate target profile estimates as shown in Fig. 7(c) and Fig. 7(d) for EOT and GPR, respectively, and listed in the row 13 of Table 2. Thus, the mKN-ML method also provides accurate target profiles for different TCP versions than that used in mKN—a useful feature for TCP versions unavailable on the Mininet host.

5.3 Exploratory Scenario Profiles

In the previous sections, the target profiles are concave, which reflect typical DTN performance with CUBIC and Hamilton TCP versions with tuned TCP parameters, and mKN profiles are convex as result of TClink implementation of connections. We now consider additional cases where in mKN profiles are non-smooth

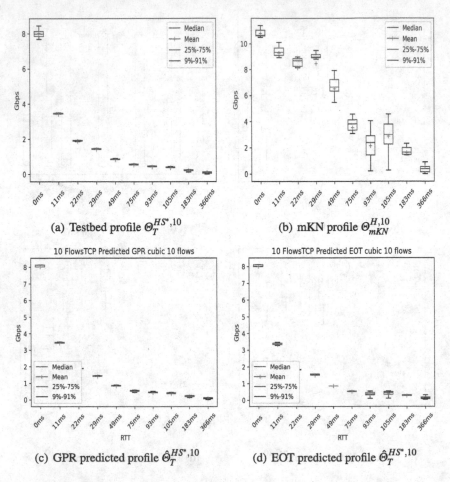

(a) Testbed profile $\Theta_T^{HS^*,10}$

(b) mKN profile $\Theta_{mKN}^{H,10}$

(c) GPR predicted profile $\hat{\Theta}_T^{HS^*,10}$

(d) EOT predicted profile $\hat{\Theta}_T^{HS^*,10}$

Fig. 9. mKN on desktop TCLink with Hamilton TCP, and testbed with HighSpeed with default TCP parameters (Table 2: row 2).

and have neither concave or convex shape. On the other hand, profiles of testbed and physical connection measurements are mostly smooth and have a dominant concave or convex shape. As shown in Fig. 8, the testbed profile is overall concave, whereas that of mKN has a much more complex shape; both GPR and EOT methods accurately estimated mostly concave profiles in this case.

When default TCP parameters are used on the testbed host, the profile is convex as shown in Fig. 9(a) for Highspeed TCP version; the profile is similarly convex for all other TCP versions we tested. Such convex profiles are also estimated by mKN method as a much more complex mKN profile is mapped to convex profile by both GPR and EOT methods, as shown in Figs. 9(c) and (d), respectively. When TCP parameters are updated to customized values for 200ms connections [18], the testbed profile becomes overall concave albeit with higher variation as shown in Fig. 10(a) for the same Highspeed TCP version. In response, concave profiles are estimated by both GPR and EOT methods.

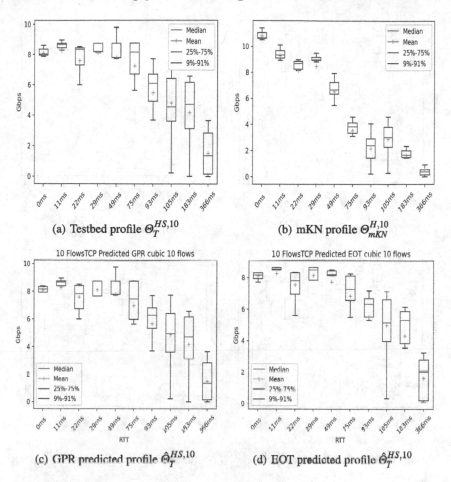

Fig. 10. mKN on desktop TCLink with Hamilton TCP, and testbed with HighSpeed with updated TCP parameters (Table 2: row 3).

Another illustrative case involves a convex mKN profile of BBR throughput measurements collected on the laptop shown in Fig. 11(b). Concave target profiles are estimated by both GPR and EOT methods, as shown in Figs. 11(c) and (d), respectively.

The VFSIE throughput profiles estimated using GPR method for four TCP versions, namely CUBIC, Hamilton TCP, Illinois and Highspeed TCP are shown in Fig. 12 (in addition to Scalable TCP profiles in Fig. 6); similarly, throughput profiles estimated by EOT method are shown in Fig. 13. These TCP versions provide similar VFSIE throughput profiles with updated TCP parameters, except for some minor variations. In general, throughput profiles for network federations emulated by Mininet can be estimated by using the mKN map computed for the host using previously collected testbed measurements for various TCP versions. These results illustrate that mKN method can be used to translate the historic

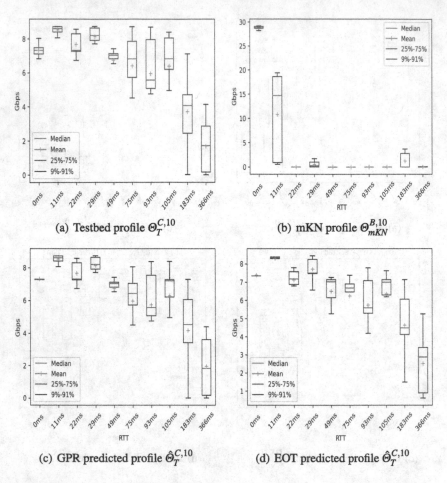

(a) Testbed profile $\Theta_T^{C,10}$

(b) mKN profile $\Theta_{mKN}^{B,10}$

(c) GPR predicted profile $\hat{\Theta}_T^{C,10}$

(d) EOT predicted profile $\hat{\Theta}_T^{C,10}$

Fig. 11. mKN on laptop with TCLink and BBR, and testbed with CUBIC (Table 2: row 8).

measurements that have been collected under various exploratory configurations to specific application scenarios.

6   Generalization Equations

In mKN-ML regression estimation, the feature $X \in \Re$ corresponds to mKN measurement and the output $Y \in \Re$ corresponds to testbed measurement at same RTT, which are jointly distributed according to $\mathbb{P}_{X,Y}$. The *expected error* of a regression function f is

$$I(f) = \int \left(f(X) - Y\right)^2 d\mathbb{P}_{X,Y}.$$

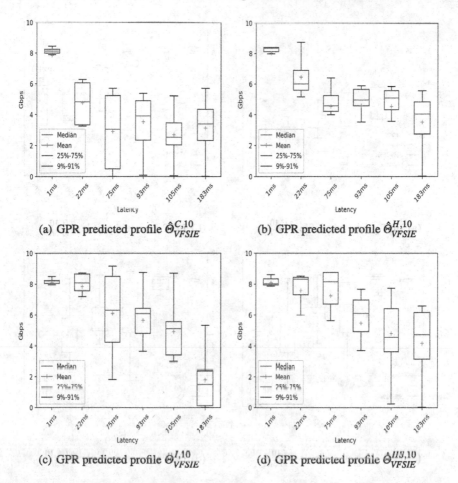

Fig. 12. Estimated VFSIE profiles using GPR method for a) CUBIC, b) Hamilton TCP, c) Illinois, and d) HighSpeed TCP versions.

The *expected best* regression estimator f^* minimizes $I(.)$ over $\mathscr{F}$, i.e., $I(f^*) = \min_{f \in \mathscr{F}} I(f)$. The *empirical error* $\hat{I}(f)$ based on training data $(X_i, Y_i), i = 1, 2, \ldots, l$, is

$$\hat{I}(f) = \frac{1}{l} \sum_{i=1}^{l} (f(X_i) - Y_i)^2.$$

It is an approximation of $I(f)$ computed based on the training data. The *empirical best* regression estimator $\tilde{f}$ minimizes $\hat{I}(.)$ over $\mathscr{F}$, i.e., $\hat{I}(\tilde{f}) = \min_{f \in \mathscr{F}} \hat{I}(f)$.

The joint distribution $\mathbb{P}_{X,Y}$ is complex, domain specific, and is only partially known. In our context, it depends on the finer details of the underlying software and hardware components, namely Mininet, VirtualBox VM and the host system, which may manifest as additional random variables. Under the boundedness

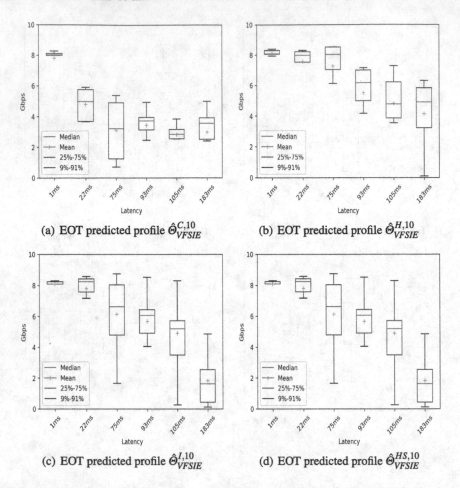

(a) EOT predicted profile $\hat{\Theta}_{VFSIE}^{C,10}$ (b) EOT predicted profile $\hat{\Theta}_{VFSIE}^{H,10}$

(c) EOT predicted profile $\hat{\Theta}_{VFSIE}^{I,10}$ (d) EOT predicted profile $\hat{\Theta}_{VFSIE}^{HS,10}$

Fig. 13. Estimated VFSIE profiles using EOT method for a) CUBIC, b) Hamilton TCP, c) Illinois, and d) HighSpeed TCP versions.

of throughput profiles, Vapnik's generalization theory [19] establishes that there exists a *confidence function* $\delta(.)$ such that for a learned map $\hat{f}$ obtained from training data we have

$$\mathbb{P}_{X,Y}^{l}\left[I(\hat{f}) - I(f^{*}) > \epsilon + \hat{\epsilon}\right] < \delta(\epsilon, l)$$

where $\epsilon > 0$ and $0 < \delta < 1$ are the precision and confidence parameters, and

$$\hat{I}(\hat{f}) \leq \min_{f \in \mathscr{F}} \hat{I}(f) + \hat{\epsilon}$$

with the training error $\hat{\epsilon}$ given by the square of RMSE in Table 2. Using mKN profile Θ_{mKN}, the estimated target profile is given by $\hat{\Theta}_T = \hat{f}(\Theta_{mKN})$. The mKN-ML regression map $\hat{f}$ corresponds to the difference between two profiles, namely mKN and testbed, both of which have bounded total variation [4], given by NC, where N is the number of RTTs and C is the connection capacity. Thus, the regressions computed by EOT and GPR methods are from a function class with total variation $V \leq 2NC$, which provides the expression

$$\delta(\epsilon, l) = 8g \left(1 + \frac{128V}{\epsilon} \right) e^{-\epsilon^2 l / 2048},$$

for a function g specified in [4]. This condition ensures that the $\hat{f}$'s error is within $\epsilon + \hat{\epsilon}$ of optimal error (of f^*) with probability $1 - \delta$. Furthermore, this probability improves with the number of measurements l, *irrespective* of the underlying joint measurement distribution $\mathbb{P}^l_{X,Y}$, which in turn depends on the randomness of mKN and testbed measurements.

7 Conclusions

We addressed the limitations of widely used Mininet network emulation tool in providing accurate throughput profiles for high bandwidth, high latency connections; in particular, these profiles can have non-representative shapes and can vary significantly across host systems with different computing and memory capabilities. We have presented a new *micro Kernel Network* (mKN-ML) method for estimating accurate throughput profiles for network testbeds and infrastructures by using (typically inaccurate) Mininet profiles from emulated high throughput, high latency networks. When used to estimate profiles for a range of network configurations, using both testbed measurements and archival data, the method achieves qualitatively similar results to direct observations, as we illustrated by presenting results for several representative examples.

Our results suggest several directions for future work. It would be of future interest to explore conditions under which Mininet emulations on more powerful servers will provide accurate throughput profiles for high bandwidth and high latency networks. It would be of future interest to explore the use of network simulators, such as ns-3, opnet, GNS3 and others, for their ability to provide accurate throughput profiles; typically, simulations take much longer computation times under these scenarios. Adaptations of the mKN approach to incorporate computing aspects into profiles by using emulated networks with vhosts representing the computing facilities would be of future interest. Future extensions of mKN approach may include: methods for shared connections with competing traffic; additional machine learning methods, including fusers of multiple, disparate (smooth, non-smooth, statistical and structural) regression methods, extensions that can handle emulation and testbed measurements collected at different RTTs; and sharpened generalization equations by taking into account finer host and network details.

Acknowledgments. This work is funded by RAMSES, SDN-SF and Applied Mathematics projects, Office of Advanced Computing Research, U.S. Department of Energy, and by Extreme Scale Systems Center, sponsored by U.S. Department of Defense, and performed at Oak Ridge National Laboratory managed by UT-Battelle, LLC for U.S. Department of Energy under Contract No. DE-AC05-00OR22725, and Argonne National Laboratory under Contract No. DE-AC02-06CH11357.

References

1. Energy sciences network. http://www.es.net
2. Experimental physics and industrial control system. epics.anl.gov
3. Lab5: Setting WAN bandwidth with token bucket filter (2019). https://bit.ly/3vhkdot
4. Anthony, M., Bartlett, P.L.: Neural Network Learning: Theoretical Foundations. Cambridge University Press, Cambridge (1999)
5. Avriel, M., Diewert, W.E., Schaible, S., Zang, I.: Generalized Concavity. SIAM, Philadelphia (2010)
6. Cardwell, N., Cheng, Y., Gunn, C.S., Yeganeh, S.H., Jacobson, V.: BBR: congestion based congestion control. ACM Queue **14**(5), 20–53 (2016)
7. Al-Najjar, A., et al.: Virtual framework for development and testing of federation software stack. In: 2021 IEEE 46th Conference on Local Computer Networks (LCN), pp. 323–326. IEEE (2021)
8. Floyd, S.: Highspeed TCP for large congestion windows. Internet draft, February 2003
9. iMars3D: Preprocessing and reconstruction for the Neutron Imaging Beam Lines. https://github.com/ornlneutronimaging/iMars3D.git
10. Kelly, T.: Scalable TCP: improving performance in high speed wide area networks. Comput. Commun. Rev. **33**(2), 83–91 (2003)
11. Phanekham, D., Nair, S., Rao, N.S.V., Truty, M.: Predicting throughput of cloud network infrastructure using neural networks. In: Workshop on Intelligent Cloud Computing and Networking (2021)
12. Rao, N.S.V., Liu, Q., Liu, Z., Kettimuthu, R., Foster, I.: Throughput analytics of data transfer infrastructures. In: Gao, H., Yin, Y., Yang, X., Miao, H. (eds.) TridentCom 2018. LNICST, vol. 270, pp. 20–40. Springer, Cham (2019). https://doi.org/10.1007/978-3-030-12971-2_2
13. Rao, N.S.V., Sen, S., Liu, Z., Kettimuthu, R., Foster, I.: Learning concave-convex profiles of data transport over dedicated connections. In: Renault, É., Mühlethaler, P., Boumerdassi, S. (eds.) MLN 2018. LNCS, vol. 11407, pp. 1–22. Springer, Cham (2019). https://doi.org/10.1007/978-3-030-19945-6_1
14. Rhee, I., Xu, L.: CUBIC: a new TCP-friendly high-speed TCP variant. In: 3rd International Workshop on Protocols for Fast Long-Distance Networks (2005)
15. Rong, R., Liu, J.: Distributed mininet with symbiosis. In: International Conference on Communications, pp. 1–6. IEEE (2017)
16. Settlemyer, B.W., Rao, N.S.V., Poole, S.W., Hodson, S.W., Hicks, S.E., Newman, P.M.: Experimental analysis of 10 gbps transfers over physical and emulated dedicated connections. In: International Conference on Computing, Networking and Communications (2012)
17. Shorten, R.N., Leith, D.J.: H-TCP: TCP for high-speed and long-distance networks. In: 3rd International Workshop on Protocols for Fast Long-Distance Networks (2004)
18. https://fasterdata.es.net/host-tuning/background/
19. Vapnik, V.N.: Statistical Learning Theory. Wiley, New York (1998)

Machine Learning Models for Malicious Traffic Detection in IoT Networks /IoT-23 Dataset/

Chibueze Victor Oha, Fathima Shakoora Farouk, Pujan Pankaj Patel, Prithvi Meka, Sowmya Nekkanti, Bhageerath Nayini, Smit Xavier Carvalho, Nisarg Desai, Manishkumar Patel, and Sergey Butakov(✉)

Concordia University of Edmonton, Edmonton, AB, Canada
{coha,ffarouk,pppatel,pmeka,snekkant,bnayini,scarvalh,ndesai,
mpatel14}@student.concordia.ab.ca, sergey.butakov@concordia.ab.ca

Abstract. Connected devices are penetrating markets with an unprecedented speed. Networks that carry Internet of Things (IoT) traffic need highly adaptable tools for traffic analysis to detect and suppress malicious agents. This has prompted researchers to explore the various benefits Machine Learning (ML) has to offer. By developing models to detect certain kinds of malicious traffic accurately, ML approach will allow for better detection capabilities if implemented in an Intrusion Detection System (IDS) or next-generation firewalls. This research paper focuses on harnessing features of ML in exploring the network traffic generated by infected IoT devices. The IoT-23 dataset was used and preprocessed into three different datasets for further exploration using various ML algorithms. This enhances the detection of malicious traffic, thereby improving the security in the IoT ecosystem. The ML algorithms implemented in this paper included: Logistic Regression, Decision Tree, Random Forest Classifier, XGBoost and Artificial Neural Network. This research was able to achieve almost 100% accuracy across all the three sub-datasets.

Keywords: Machine Learning · IoT-23 · Internet of Things · Supervised learning · TensorFlow

1 Introduction

The need for more effective and efficient techniques for detecting malicious connections in various networks systems will keep growing with an unprecedented speed in our hyperconnected world. This trend has prompted researchers to explore the various benefits machine learning (ML) has to offer. In 2019, CIRA conducted a survey which revealed that 71% of organizations in Canada experienced at least one form of cyber-attack which led to losses in terms of finances, time, and resources. This costs Canadian organizations an average of $9.25 million in investigation and remediation [1] and clearly shows how important it is to put network security into consideration by investing in advanced detection technologies. The same applies to other developed and developing countries since the number of connected devices is growing exponentially across the globe.

© Springer Nature Switzerland AG 2022
É. Renault et al. (Eds.): MLN 2021, LNCS 13175, pp. 69–84, 2022.
https://doi.org/10.1007/978-3-030-98978-1_5

Internet of Things (IoT) refers to physical devices that can interconnect with other devices in a network and exchange information to perform some functionality for the user. Market of IoT devices have grown exponentially over the last few years. These devices have use in healthcare, industrial applications and consumer areas with devices like smartwatches, home assistants and smart TVs. In research published by Transformation Insights [2], they estimated the number of active IoT devices were over 7.6 billion at the end of 2019. They further estimate that this figure will grow to 24.1 billion IoT devices by 2030.

Rush to put IoT devices on the market clearly led to a common issue with IoT devices: many of them are not built with security in mind and thus are susceptible to attacks that can compromise the network, information or other devices. Some common issues with IoT devices are that they are poorly configured and firmware updates and patches are not regularly provided and pushed out, which led to unpatched vulnerabilities. Some of the possible attacks on IoT devices have been documented as follows:

- Physical Attacks: Physical attacks on IoT devices target the hardware of an IoT device and by interfering with the device's hardware components, a malicious actor may gain control of the device. These attacks are carried out while the attacker is in close proximity to a network or an IoT device. Some examples of physical attacks on IoT devices include, Node tampering, Radio frequency (RF) Interference, Node attack, Node jamming, Physical damage and Social engineering attacks.
- Software Attacks: Most IoT devices run some kind of application on top of the operating system (OS) to perform functionality for users. An IoT device can become compromised on a software level by conducting phishing attacks, introducing malware, spyware or a backdoor into the device. A compromised IoT device can also be used to run malicious code, cause a buffer overflow or conduct side-channel attacks to extract sensitive data from protected memory space.
- Network Attacks: Network attacks refer to the possible attacks that may occur during the transmission of data between devices in a network. If a compromised IoT device is present in a network it could cause attacks such as eavesdropping, man-in-the-middle attacks, Denial of Service (DOS) or Distributed Denial of Service (DDOS) by infecting the IoT device with a botnet malware. There have also been attacks documented on the routing protocol for low-power and Lossy networks which include the Sybill Attack, Selective Forwarding, Wormhole attack, Sinkhole Attack, Blackhole Attack and Hello flooding attack [3].

Given the fact that IoT ecosystems consist of heterogeneous devices, they require flexible and adaptable tools to protect network communications between them. Approaches in detecting malicious traffic using ML have increased exigently over the past few years with several techniques proposed with the sole purpose of enhancing malicious traffic detection [4]. Machines learn by experience from models designed by humans to analyze data input to make accurate predictions. Additionally, the advancement of Artificial intelligence (AI) and ML techniques along with the decreased cost of computing power have helped researchers and security professionals to implement these techniques into Intrusion Detection Systems (IDS). These IDS serve as a means of detecting malicious traffic and have only grown more sophisticated in recent years.

The proposed research looks at the approaches of equipping security applications with ML-based tools to protect IoT networks.

The sections of this research paper are structured as follows: Section 2 gives an overview of the traditional methods that were used in network traffic analysis and the existing ML approaches that can be employed in detecting malicious IoT traffic. Section 3 provides a detailed description of the IoT-23 dataset used and how it was engineered for further analysis. Section 4 discusses the preprocessing stages of the IoT-23 dataset used. Section 6 outlines the experimentational result of analyzing the IoT traffic using ML models identified in Sect. 5. Section 7 compares the result gotten to other similar research papers, detailing their differences. Finally, the Conclusion provides a summary of the research, outlines future improvements and practical implementations in industrial environment.

2 Review of the Related Works

Traffic analysis is one of the major tools that is being used to detect and isolate infected hosts. In the following review the methods for traffic analysis are split into two groups: traditional methods and methods that involve ML. Although ML methods are being used for traffic analysis for years, they are yet to make it to mainstream market and still can be considered as new to the IoT ecosystems.

2.1 Traditional Methods for Network Traffic Analysis

The most common network traffic analysis methods involved the use of traffic outlier detection algorithms and the use of Intrusion Detection Systems (IDS).

1) Traffic Outlier Detection

An outlier is a data point that is expressively different from other normal remarks [5]. There are multiple algorithms to identify the outliers in the urban network traffic. They include *Flow outlier detection* and *Trajectory outlier detection*.

Flow based outlier detection is used to find anomalies by inspecting the header information carried out by the flow analyzers [6]. The methods used to analyze the network traffic are statistical, ML, clustering, frequent pattern mining and agent based [7]. The use of trajectory outlier detection is to learn trajectories or their sections which alter noticeably from or are unreliable with the residual set. The trajectory outliers include offline processing and online processing [7].

Intrusion Detection Systems

The unauthorized access to an information system can be referred as an intrusion. In short, any kind of threat that can affect the enterprise's information confidentiality, integrity and availability is an intrusion [8]. The software applications or devices were developed to detect the intrusion to the system which is called an intrusion detection system (IDS). The purpose of the IDS is not only to prevent the attack but to identify and report it to the network administrator [9]. There are two learning techniques according to which intrusion detection system behaves:

Signature based Intrusion Detection System
Signature-based intrusion detection system uses known patterns or a signature of the malicious traffic to identify the attack traffic. The known patterns are stored in a database which includes the collection of the suspicious activities and operations that can exploit the weaknesses of the information systems. In this technique, the pattern of the incoming traffic is compared with the pattern stored in the database to differentiate the attack traffic from the legitimate traffic [8]. The SNORT tool is a great example of a signature-based intrusion detection [9].

Anomaly intrusion detection system
To develop an anomaly-based intrusion detection system, the baseline for the network traffic needs to be decided. The deviation of network traffic behavior from the baseline is considered an intrusion. The behavior of the attack traffic which is not like the legitimate traffic can be treated as the intrusion [10]. This type of system can be attached to either network-based or host-based intrusion detection system.

The fundamental edge of the anomaly-based intrusion detection system is its ability in detecting unknown attacks and can be treated as the best solution against zero-day attacks. It is very difficult for the attackers to discover what is the normal behavior decided by the system [8]. Anomaly based intrusion detection system can be categorized in three types as per the training process as follows: (a) Statistics based, (b) Knowledge-based, and (c) ML based.

Conventional Packet Inspection
Conventional packet filtering reads only packet header information which is called a stateful packet inspection. This is not a sophisticated way to filter the packets as it does not investigate the data part (payload) of the packet [11]. The stateful packet inspection can be done by using firewalls. The disadvantages of stateful packet inspection can be overcome by using deep packet inspection. *The research represented in this paper can be classified as conventional packet inspection with the use of ML methods.*

Deep Packet Inspection
Deep packet inspection is also called DPI. It is an information extraction or packet inspection method. Deep packet inspection carries out an inspection into the information from header of the packets and the data part of the packet at a particular examination point. It checks for the specified protocol, spam, viruses, intrusions and any other defined malicious factor to deny the packet from passing over the examination point [11]. Deep packet inspection makes decisions about whether a specific packet should be dropped or forwarded to the destination.

2.2 Approaches to Detecting IoT Malicious Traffic Using Machine Learning

The various approaches to malicious traffic detection can be grouped into three categories which include supervised learning, unsupervised learning, and reinforcement learning.

Supervised Learning
In the Supervised Learning model, the algorithm is given a completely labelled dataset

that it can use to test the accuracy of training data. The model uses the training dataset to learn and creates its own logic to determine that the right outcome is achieved. The testing data set is then fed to the model which can test the model to see how well it learnt during the training phase. [12]. Supervised learning can be categorized as follows: (a) Classification which can be either binary or multi-class. Multiclass, multi groups can be expected in this sort of classification model. Binary, unless this is predicted by a Boolean value, i.e., 0 or 1, or whether the value is true or false, as in the multielass classification form [12, 13]; (b) Regression–when the output variable is a real number, regression is used. Some examples of regression methods include Linear Regression, Random Forest and Support vector machines (SVM).

Unsupervised Learning
In unsupervised learning, the computer is trained with knowledge that is neither classified nor numbered. The algorithm then attempts to group the unsorted data by extracting useful features based on similarities, patterns, and differences [14]. Clustering and dimension reduction problems are two subtypes of unsupervised learning problems.

Reinforcement Learning
In reinforcement learning, the aim is to continually observe the environment and use the knowledge gained to improve upon the model. The model works towards the final goal this way by trial and error through observation of the surrounding environment [15].

All the ML methods mentioned above require adequate datasets that reflect the traffic in real IoT networks. Various research groups attempted to create good IoT datasets in order to provide common ground for researchers to test or sharpen ML models for malicious traffic detection in IoT networks. Examples of such datasets include the IoT-23 dataset, CGIAR dataset, TLESS dataset, etc. This research looked at one of the most recent traffic datasets generated by clean and infected IoT devices: IoT-23. The dataset represents wide variety of the traffic patterns and attacks and provides sufficient amount of data suitable for most ML algorithms [16, 17].

3 IoT 23 Dataset

3.1 Description of IoT-23 Dataset

IoT-23 is a recent dataset which comprises of network traffic acquired from IoT devices. It encompasses twenty-three captures (20 malwares and 3 benign traffic) all captured within the year 2018 to 2019 by Avast AIC laboratory in partnership with the Czech Technical University in Prague. The IoT-23 dataset provides a large data source of properly labelled real malware and benign IoT traffic for ML research purposes. Traffic was generated from three hardware IoT devices namely, Amazon Echo Home device, Philips HUE smart LED lamp and a Somfy smart door lock. The generated traffic included the following upper layer protocols: HTTP, SSL, DBS, DHCP, Telnet and IRC. The dataset has a capture of 764,735,276 packers with 764,308,000 being malicious in nature. Table 1 below shows a summary of the types of labels contained in the IoT-23 data [16].

Table 1. Summary of labels in IoT-23 Dataset

Labels	Description
Attack	Malicious attack packets from an infected host to another host
Benign	Genuine packets
C&C	Infected devices had connections to a C&C server
DDoS	DDoS attacks carried out on infected devices are indicated with this label
HeartBeat	Packets sent from the suspicious source to keep alive the connection on the infected host by the C&C server
FileDownload	Files downloaded to the infected devices are indicated with this label
Mirai	Data have similarities and features of a Mirai Bonet
Okiru	Data have similarities and features of an Okiru Bonet
PartOfAHorizontalPortScan	Information gathered through horizontal port scan for further attacks
Torii	Data have similarities and features of a Torii Bonet

The IoT-23 dataset comprises of conn_log_labelled files each containing 23 columns of data. A detailed description of the 23 columns can be found on the IoT-23 dataset website [16].

3.2 Data Engineering

Among 23 sub datasets in IoT-23 dataset, some of them are very small in size i.e., in kilobytes and some are larger in size i.e., more than 2 GB. Also, all 23 datasets in IoT-23 are severely imbalanced. 12 malicious labels and a benign label are identified in all 23 datasets. Among these 12 malicious labels, Malicious PartOfHorizontalPortScan, Malicious C&C Okiru and Malicious DDoS labels represent vast majority of labels (millions), and rest of the malicious labels are much less in number. All these labels are not present in every dataset. The detailed description of each sub-dataset is provided in [17]. So, it has been decided to create one file for one label such that all the data of that label is taken from the datasets containing that label and grouped into the file. So, in total 13 files are created for the 13 labels containing the data from all the 23 datasets.

To balance the dataset and overcome the issue of underfitting and overfitting on the IoT-23 datasets, 3 datasets were created by taking random amount data from all the 13 files. In this process, randomness in data selection was always maintained. Figures 1, 2 and 3 present compositions for the three datasets.

For creating 'Dataset_1', N random data was selected from each of the 12 malicious files and mixed with the benign entries. It contains all 12 malicious labels and benign labels. In this dataset, benign labels are encoded as 1 and all malicious labels are encoded as 0.

For creating 'Dataset_2', N random data was selected from 3 most common labels for malicious traffic and mixed with the benign entries in a balanced way. This dataset contains Benign, Malicious DDoS, Malicious PartOfAHorizontalPortScan and Malicious Okiru labels.

For creating 'Dataset_3', the other malicious labels represented in much smaller quantities had been taken randomly and mixed with the benign label. An oversampling technique was used to make the dataset balanced. This dataset includes 10 different types of labels. The labels taken in this dataset are Benign, Malicious C&C-FileDownload, MaliciousC&C, Malicious C&C-Mirai Malicious FileDownload, Malicious Attack, Malicious C&C-HeartBeat, Malicious Torii, Malicious C&C-PartofAHorizontalPortscan, and Malicious C&C-HeartBeat-FileDownload. In feeding the engineered dataset to the different machine learning models, the dataset labels were encoded as Benign traffic with the encoding of 1 and various Malicious traffic was encoded from 2 to 12.

4 Preprocessing

Dropping of Non-unique and Unimportant Columns: Preliminary analysis of the datasets indicated that columns 'local_orig' and 'local_resp' are non-unique. So, these columns had been dropped along with 'u_id' column as it does not have any importance in building a model.

Converting Categorical Columns to Numerical Datatype: Columns like 'id_orig_h', 'id_resp_h', 'proto', 'service', 'conn_state', 'history' are categorical, and are therefore had been converted to numerical datatype using Label Encoder.

Checking for Missing Values: The columns 'duration', 'orig_bytes' and 'resp_bytes' contained missing values. These missing values were replaced by mean value of their respective column.

Splitting of data into Train and Test data: After the above steps were completed, the dataset was split into train and test datasets in the ratio of 7:3.

Feature Scaling: Feature scaling limits the data variables of each column to certain range to be compared on common grounds. Standard Scaler technique was implemented on the datasets to standardize the range of data with zero mean and standard deviation of one.

5 Models

To build classification engines on the prepared datasets, five machine learning models have been used: Logistic Regression, Decision Tree Classifier, Random Forest Classifier, XGBoost Classifier and Artificial Neural Network. Tensorflow Python library was used to build all models. The source code for all the models is available at [17].

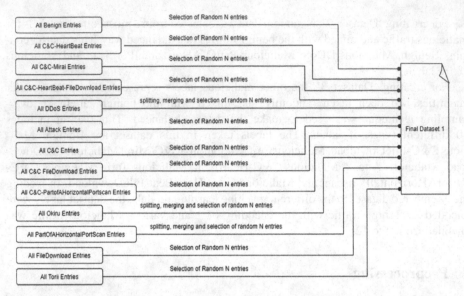

Fig. 1. Dataset preparation for Dataset_1

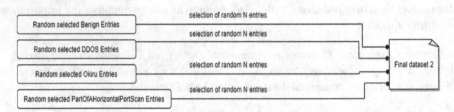

Fig. 2. Dataset preparation for Dataset_2

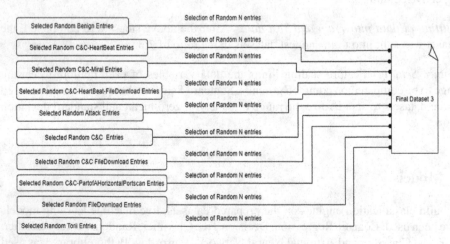

Fig. 3. Dataset preparation for Dataset_3

5.1 Decision Tree Classifier

Decision tree models for Dataset_1 and Dataset_2 have been configured with default settings. While Dataset_3 decision tree model parameters have been configured with Gini split criterion, and minimum and mean sample leaf size to be 1 and 2 respectively.

5.2 Logistic Regression

Logistic regression model used on each dataset consist of two hidden layers. In Dataset_1, first hidden layer's weight TensorFlow variable is calculated using shape as (17, 12) and biases TensorFlow variable with shape as (12,). Second hidden layer's weight Tensor-Flow variable is calculated using shape as (12, 10) and biases TensorFlow variable with shape as (10,). In Dataset_2, first hidden layer's weight TensorFlow variable is calculated using shape as (17, 8) and biases TensorFlow variable with shape as (8,). Second hidden layer's weight TensorFlow variable is calculated using shape as (8, 4) and biases TensorFlow variable with shape as (4,). Lastly on Dataset_3, first hidden layer's weight TensorFlow variable was calculated using shape as (17, 4) and biases Tensor-Flow variable with shape as (4,). Second hidden layer's weight TensorFlow variable was calculated using shape as (4, 2) and biases TensorFlow variable with shape as (2,). All logistic regression model's first hidden layer has rectified linear activation function (Relu).

5.3 Random Forest Classifier

Random Forest Classifier model of all three datasets had been configured default parameters set in the TensorFlow library.

5.4 XGBoost Classifier

XGBoost classifier model in all three datasets was configured with different random state and learning rate values. For Dataset_1 and Dataset_2, this model's values were set with random state as 1 and learning rate as 0.01. While for Dataset_3, this model's values were set to random state as 42 and learning rate as 0.1.

5.5 Artificial Neural Networks

ANN sequential model for all 3 datasets were configured with 1 input layer, 1 hidden layer and 1 output layer. First hidden layer of all models has (17, null) as input shape, and rectified linear activation function. The output layer of all models for Dataset_1, Dataset_2 and Dataset_3 was 10,4 and 2 respectively. All models for three of the different datasets were compiled with optimizer implementing Adam algorithm, loss function set as categorical cross entropy, and accuracy used as learning metrics.

6 Results

To determine the effectiveness of the ML algorithms employed in this research, metrics such as accuracy, precision, recall, AUC_ROC and F1 score were used. Accuracy gives an overview of how effective the ML algorithms are in detecting malicious and non-malicious traffic. It demonstrates the algorithm's ability to differentiate false positives and false negative. F1 score helps to determine the precision of the classifier. A higher F1 score indicates a better performance of the analysis. The number of true positives to the total number of real positives is expressed as Recall.

Dataset_1 contains random data selected from each of the 12 malicious files and mixed with benign entries. It contains all 12 malicious labels and benign labels. Results from analysis of Dataset_1 are shown in Table 2. As shown in Table 2 for Dataset_1, Logistic regression produced the least accuracy of 96.1% and F1 score of 0.96. On the other hand, artificial neural network and Random Forest classifier produces a much better accuracy of 99.99% and F1 score of 1 because of their ability to evaluate complex inputs.

A Confusion matrix provides a summary of the performance of classification algorithm when evaluating a machine learning classification problem. It is the ratio of accurate prediction to the overall prediction made. The confusion matrix diagram for dataset_1 is shown in Table 3.

Table 2. Result analysis summary of Dataset I

Algorithm	Evaluation metrics			
	Accuracy	Precision	Recall	F1 score
Logistic regression	96.1%	0.93	0.99	0.96
Random forest classifier	99.99%	1	1	1
Decision tree classifier	99.99%	1	1	1
XGBoost classifier	99.90%	0.99	0.99	0.99
Artificial neural network	99.99%	1	1	1

Table 3. Confusion Matrix summary of Dataset_1

		Predicted Label			
		0		1	
Actual Label	0	92.71%	99.99%	7.29%	0.01%
		99.99%	99.98%	0.01%	0.02%
		99.79%		0.21%	
	1	0.57%	0.01%	99.43%	99.99%
		0.01%	0.17%	99.99%	99.83%
		0.19%		99.81%	

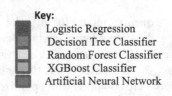

Key:
Logistic Regression
Decision Tree Classifier
Random Forest Classifier
XGBoost Classifier
Artificial Neural Network

Confusion matrixes of Dataset_1 shows that Logistic regression had the least performance due to a higher number of false positives of 0.57% and false negative of 7.29% when compared to other machine learning algorithms. Nevertheless, Decision Tree classifier had the best performance with a lower false positive and false negative of 0.01%.

In Dataset_2, random data was selected from 3 large malicious labels files and mixed with the benign entries in a balanced way. Results from the analysis of Dataset_2 are as shown in Table 4.

Like the analysis of Dataset_1, logistic regression produced lesser accuracy of 98.9% compared to Random Forest classifier, Decision Tree Classifier and Artificial Neural Network due to high false positives and false negatives when analyzing Dataset_2.

For the confusion matrixes of Dataset_2, Logistic Regression Analyses had a lower performance with higher false negatives and false positives compared to Decision Tree Classifier and Artificial Neural Network which performed much better with more precise predictions (Table 5).

Table 4. Result analysis summary of Dataset II

Algorithm	Evaluation metrics			
	Accuracy	Precision	Recall	F1 score
Logistic regression	98.8%	0.989	0.988	0.988
Radom forest classifier	99.99%	1	1	1
Decision tree classifier	99.99%	1	1	1
XGBoost classifier	99.99%	0.99	0.99	0.99
Artificial neural network	99.99%	1	1	1

Table 5. Confusion Matrix Summary of Dataset_2

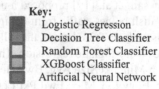

Key:
Logistic Regression
Decision Tree Classifier
Random Forest Classifier
XGBoost Classifier
Artificial Neural Network

Dataset_3 contains the other malicious labels which are small. They are taken randomly and mixed with the benign label. An oversampling technique was used to make the dataset balanced. This dataset includes 10 different types of labels. Results from analysis of Dataset_3 are shown in Table 6.

Table 6. Result Analysis Summary of Dataset III

Algorithm	Evaluation metrics			
	Accuracy	Precision	Recall	F1 score
Logistic regression	98.1%	0.982	0.981	0.981
Random forest classifier	99.98%	0.99	0.99	0.99
Decision tree classifier	99.98%	0.99	0.99	0.99
XGBoost classifier	99.95%	0.99	0.99	0.99
Artificial neural network	99.98%	0.99	0.99	0.99

Confusion matrixes of Dataset_3 shows that Radom Forest classifier algorithm had a better performance with the least number of false positives and false negatives when compared to other algorithms used (Table 7).

7 Discussion

Researchers in [18] looked at the applicability of ML algorithms in the identification of anomalies in IoT network traffic. They compared different machine learning algorithms against the IoT-23 dataset which contains both malicious and benign traffic. The following machine learning algorithms were used in the comparison: Random Forest (RF), Naïve Bayes (NB), Multi-Layer Perceptron, a variant of the Artificial Neural Network (ANN), Support Vector Machine (SVM) and AdaBoost (ADA). As per the results of the research, the Random Forest algorithm returned 100% accuracy and was deemed the best algorithm for this dataset (Table 8).

Hussain et al. [19] proposed a universal feature set for machine learning models to distinguish botnet traffic from benign traffic regardless of the underlying dataset. With regards to the IoT-23 dataset, the following features were selected as the 'most universal' features using logistic regression. They are as follows; Pkt Len Mean, Bwd Pkt Len Min, Pkt Len Min, Pkt Size Avg, Bwd Header Len, Bwd IAT Max, Bwd Pkt Len Mean, Flow Byts/s, Flow IAT Max, Fwd Pkt Len Mean. Four ML algorithms were used to classify the traffic, they are: Naïve Bayes, K nearest neighbor, Random Forest, and Logistic Regression. Of the four, Random Forest performed the best, but all other algorithms used returned above 98% in accuracy, precision, recall and F1- score. When article [19] is compared with this research, this research performed better with a percentage of 99.9% because of how the IoT-23 dataset was engineered.

In research conducted by Kumar et al., [20] they proposed and developed EDIMA, a lightweight IoT botnet detection solution deployed at edge gateways which provides early detection of botnets before they can coordinate an attack. EDIMA consists of a two-stage detection mechanism. The first uses Machine Learning techniques to classify

Table 7. Confusion matrix summary of Dataset_3.

Logistic Regression

Actual Label \ Predicted Label	0	1	2	3	4	5	6	7	8	9
0	99.51%	0	0.04%	0.17%	0	0.08%	0.13%	0	0.04%	0
1	0	99.06%	0	0	0	0	0	0	0	7.3%
2	0.23%	0	90.80%	0	0	0.64%	0	0%	7.90%	0
3	0	0	0	100%	0	0	0	0	0	0
4	0	0	0	0	100%	0	0	0	0	0
5	0	0	1.87%	0.04%	0	97.99%	0	0	0.09%	0
6	0	0	0.04%	0	0	0	99.96%	0	0	0
7	0	0	0	0	0	0	0	100%	0	0
8	0	0	0	0	0	0	0	0	100%	0
9	0	0	0	0	0	0	0	0	0	100%

Decision Tree Classifier

Actual Label \ Predicted Label	0	1	2	3	4	5	6	7	8	9
0	100%	0	0	0	0	0	0	0	0	0
1	0	100%	0	0	0	0	0	0	0	0
2	0.08%	0.04%	99.88%	0	0	0	0	0	0	0
3	0	0	0	100%	0	0	0	0	0	0
4	0	0	0	0	100%	0	0	0	0	0
5	0	0	0	0	0.04%	99.92%	0	0	0.04%	0
6	0	0	0	0	0	0	100%	0	0	0
7	0	0	0	0	0	0	0	100%	0	0
8	0	0	0	0	0	0	0	0	100%	0
9	0	0	0	0	0	0	0	0	0	100%

Random Forest Classifier

Actual Label \ Predicted Label	0	1	2	3	4	5	6	7	8	9
0	99.92%	0	0.08%	0	0	0	0	0	0	0
1	0	100%	0	0	0	0	0	0	0	0
2	0	0	100%	0	0	0	0	0	0	0
3	0	0	0	100%	0	0	0	0	0	0
4	0	0	0	0	100%	0	0	0	0	0
5	0	0	0	0	0.04%	99.96%	0	0	0	0
6	0	0	0	0	0	0	100%	0	0	0
7	0	0	0	0	0	0	0	100%	0	0
8	0	0	0	0	0	0	0	0	100%	0
9	0	0	0	0	0	0	0	0	0	100%

XGBoost Classifier

Actual Label \ Predicted Label	0	1	2	3	4	5	6	7	8	9
0	99.81%	0	0.08%	0	0	0.08%	0.03%	0	0	0
1	0	100%	0	0	0	0	0	0	0	0
2	0	0	99.85%	0	0	0.15%	0	0	0	0
3	0	0	0	100%	0	0	0	0	0	0
4	0	0	0	0	100%	0	0	0	0	0
5	0	0	0.04%	0	0.04%	99.84%	0	0.04%	0.04%	0
6	0	0	0	0	0	0	100%	0	0	0
7	0	0	0	0	0	0	0	100%	0	0
8	0	0	0	0	0	0	0	0	100%	0
9	0	0	0	0	0	0	0	0	0	100%

Artificial Neural Network

Actual Label \ Predicted Label	0	1	2	3	4	5	6	7	8	9
0	99.88%	0	0	0.04%	0	0.04%	0	0	0.04%	0
1	0	100%	0	0	0	0	0	0	0	0
2	0	0	99.07%	0	0	0	0.85%	0	0	0.08%
3	0	0	0	100%	0	0	0	0	0	0
4	0	0	0	0	100%	0	0	0	0	0
5	0	0	0.07%	0.04%	0	99.89%	0	0	0	0
6	0	0	0	0	0	0	100%	0	0	0
7	0	0	0	0	0	0	0	100%	0	0
8	0	0	0	0	0	0	0	0	100%	0
9	0	0	0	0	0	0	0	0	0	100%

Legend:
- Logistic Regression
- Decision Tree Classifier
- Random Forest Classifier
- XGBoost Classifier
- Artificial Neural Network

Table 8. The results from various methods from Stoian [18]

Matrices		Classifiers				
		RF	NB	ANN	SVM	ADA
Precision	Weighted	1.00	0.76	0.71	0.60	0.86
	Macro	0.88	0.27	0.33	0.23	0.55
Recall	Weighted	1.00	0.23	0.66	0.67	0.87
	Macro	0.85	0.38	0.14	0.14	0.35
F1_score	Weighted	1.00	0.25	0.52	0.59	0.83
	Macro	0.84	0.10	0.10	0.13	0.37
	Accuracy	1.00	0.23	0.66	0.67	0.87

aggregate network traffic and the second stage uses ACF based tests to detect individual bots. Three types of ML algorithms were used in this experiment, Gaussian Naive Bayes', SVM and Random Forest. Again, the Random Forest algorithm performed the best and had the highest accuracy, recall, precision and F1 score out of the three algorithms used.

As it can be seen from the confusion matrixes in the Results section, most of mis-classifications are happening between various types of attacks. It essentially means that for class 0 –benign traffic–there was no false negatives. False negatives in this case allow malicious traffic to stay undetected. Cases where benign traffic was classified as malicious are also below 1% so that means that false positives will not be triggering intrusion alarms at the higher rate. The reasons for the false positives need to be studied further for the applications that are sensitive to single packet failures.

Although the engineering and preprocessing of the IoT-23 dataset implemented here differ from other research papers compared above, Random Forest classifier also performed better in anomaly detection and classification. As described in above Section III, 3 different datasets were created namely Dataset_1, Dataset_2 and Dataset_3 from the IoT-23 dataset. Moreover, the IoT-23 dataset was too big in size and some of the attack categories were small, which would easily result into overfitting of models. So, dividing this big dataset into three smaller datasets with different categories has presented a chance to train models with all the attack labels while maintaining memory and CPU requirements at the reasonable level that is typical to standalone traffic analysis devices.

Practically, the proposed ML models can be used to build traffic analysis modules for intrusion detection systems for networks that carry traffic from IoT devices. It can be done in two main phases namely detection and classification. Detection phase is used to extract the features and its value from the packets of IoT network. Specifically, it looks at the features and its value which the machine learning model is trained for. It puts all the information of the packets in the form of a record. Secondly, the classification phase can be used to classify the data records obtained in the pre-processed phase. This phase recognizes that the identified packet to be benign or malicious.

As IoT devices continue to become more popular in corporate and home environments, there is a growing need to maintain the security of the network these devices are connected to. Specific malware that targets vulnerable IoT devices can create a point of weakness and compromise the security of the network. As a result, designing and implementing such ML models and integrating them with IDS devices will allow for better detection of malicious IoT traffic while providing high level of adaptability for intrusion detection applications that can evolve along with the evaluation of IoT devices on the networks they monitor.

The advantage of this approach is that it gives an opportunity to train all models in a similar way to produce better performance matrix results of all three datasets with all labels. In addition, less computational power is required in training all three sub datasets as compared to training the full dataset at once. In real time, it can give good detection capability for all kinds of attacks which will make detection more accurate, and ultimately making IoT ecosystems more secured.

8 Conclusion

Deep examination of IoT-23 datasets allowed to observe that most of the datasets are imbalanced. Significant data preprocessing was conducted to prepare three final datasets out of the 23 separate datasets included in the original IoT-23 dataset. Various machine learning models were implemented using these final datasets but only five models, namely Logistic Regression, Decision Tree Classifier, Random Forest Classifier, XG Boost Classifier and Artificial Neural Network are considered for this research based on the time taken to train the models. Among these five models, the time taken to train Decision tree model and Artificial Neural network model is less when compared to the Logistic Regression, Random Forest and XG Boost models. In terms of model evaluation metrics, except for Logistic Regression, the other four model's performance is almost 100% across the three datasets.

To improve the models, further research can be done for the security of IoT device which should be conducted in the post-processing phase. An algorithm can be developed which can further classify the data in groups of records for the second time to reduce the false ratio of the proposed models. Possible future development work would be to implement these trained ML models in an IDS device or a next generation firewall in an IoT network and monitor how well the models perform in a real-world environment.

Transformed datasets and scripts used in this research can be found at [19].

References

1. CIRA: 2019 CIRA Cybersecurity Survey, CIRA (2019). https://www.cira.ca/resources/cyb ersecurity/report/2019-cira-cybersecurity-survey. Accessed 2 March 2021
2. Transforma Insights: Global IoT Market Will Grow to 24.1 Billion Devices in 2030, Generating $1.5 Trillion Annual Revenue, CISON PR Newswire (2020). https://www.prn ewswire.com/news-releases/global-iot-market-will-grow-to-24-1-billion-devices-in-2030-- generating-1-5-trillion-annual-revenue-301061873.html. Accessed 20 March 2021
3. Khraisat, A., Alazab, A.: A critical review of intrusion detection systems in the internet of things: techniques, deployment strategy, validation strategy, attacks, public datasets and challenges. Cybersecurity 4(18) (2021)

4. Beigi, E.B., Jazi, H.H., Stakhanova, N., Ghorbani, A.A.: Towards effective feature selection in machine learning-based botnet detection approaches. In: 2014 IEEE Conference on Communications and Network Security, San Francisco, CA, USA, pp. 247–255 (2014). https://doi.org/10.1109/CNS.2014.6997492
5. Jae-Gil, L., Jiawei, H., Xiaolei, L.: Trajectory outlier detection:a partition-and-detect framework. In: 2008 IEEE 24th International Conference on Data Engineering, Cancun, Mexico (2008)
6. Sharma, R., Guleria, A., Singla, R. K.: An overview of flow-based anomaly detection. Int. J. Commun. Netw. Distrib. Syst. **21**, 220–240 (2018) https://doi.org/10.1504/IJCNDS.2018.10014505
7. Djenouri, Y., Belhadi, A., Lin, J.C.-W., Djenouri, D., Cano, A.: A survey on urban traffic anomalies detection algorithms. IEEE Access **7**(2169–3536), 12192–12205 (2019)
8. Khraisat, A., Gondal, I., Vamplew, P., et al.: Survey of intrusion detection systems: techniques, datasets and challenges (2019). https://doi.org/10.1186/s42400-019-0038-7. Accessed 10 October 2020
9. Asif, M.K., Khan, T.A., Taj, T.A., Naeem, U.: Network intrusion detection and its strategic importance. In: IEEE Business Engineering and Industrial Applications Colloquium (BEIAC) (2013). https://doi.org/10.1109/beiac.2013.6560100
10. Jyothsna, V., Prasad, K.M.: Anomaly-based intrusion detection system (2019). https://doi.org/10.5772/intechopen.82287
11. Brook, C.:What is Deep Packet Inspection? How It Works, Use Cases for DPI, and More (2018). https://digitalguardian.com/blog/what-deep-packet-inspection-how-it-works-use-cases-dpi-and-more. Accessed 15 November 2020
12. Data Robot: Supervised Machine Learning. https://www.datarobot.com/wiki/supervised-machine-learning/. Accessed 28 October 2020
13. Brownlee, J.: Supervised and Unsupervised Machine Learning Algorithms (2020). https://machinelearningmastery.com/supervised-and-unsupervised-machine-learning-algorithms/. Accessed 28 October 2020
14. Shai, S., Shai, B.: Understanding Machine Learning: From Theory to Algorithm, Cambridge University Press (2014)
15. Ansam, K., Ammar, A.: A critical review of intrusion detection, pp. 1–27 (2021). https://doi.org/10.1186/s42400-021-00077-7
16. Parmisano, A., Garcia, S., Erquiaga, M. J.: Stratosphere Laboratory. A labeled dataset with malicious and benign IoT network traffic. https://www.stratosphereips.org/datasets-iot23
17. GitHub: Machine-Learning-Models-for-Detecting-Malicious-Traffic-in-IoT-Devices-using-IoT-23-Dataset. https://github.com/Bhageerath123/Machine-Learning-Models-for-Detecting-Malicious-Traffic-in-IoT-Devices-using-IoT-23-Dataset. Accessed 20 March 2021
18. Stoian, N.A.: Machine Learning for Anomaly Detection in IoT networks: Malware analysis on the IoT-23 Data set. University of Twente, Enschede, Netherlands (2020)
19. Hussain, F., Abbas, S.G., Fayyaz, U.U., Shah, G. A., Toqeer, A., Ali, A.: Towards a universal features set for IoT botnet attacks detection. In: IEEE 23rd International Multitopic Conference (INMIC) (2020)
20. Kumar, A., Shridhar, M., Swaminathan, S., Lim, T.J.: Machine Learning-Based Early Detection of IoT Botnets Using Network-Edge Traffic Singapore. University of Technology and Design, Singapore (2020)

Application and Mitigation of the Evasion Attack against a Deep Learning Based IDS for IoT

Nicholas Lurski[1,2]([✉])[iD] and Mohamed Younis[2][iD]

[1] Johns Hopkins University Applied Physics Laboratory, Laurel, MD 20723, USA
Nicholas.Lurski@jhuapl.edu
[2] University of Maryland, Baltimore County, Baltimore, Maryland 21250, USA
younis@umbc.edu

Abstract. An Internet of Things (IoT) network is characterized by ad-hoc connectivity and varying traffic patterns where the routing topology evolves over time to account for mobility. In an IoT network, there can be an overwhelming number of massively connected devices, all of which must be able to communicate to each other with low latency to provide a positive user experience. Various protocols exist to allow for this connectivity, and are vulnerable to attack due to their simple nature. These attacks seek to disrupt or deny communications in the network by taking advantage of these vulnerabilities. These attacks include Blackhole, Grayhole, Flooding and Scheduling attacks. Intrusion Detection Systems (IDS) to prevent these routing attacks exist, and have begun to incorporate Deep Learning (DL) to bring near perfect accuracy of detection of attackers. The DL approach opens up the IDS to the possibility of being the victim of an Adversarial Machine Learning attack. We explore the case of a novel evasion attack applied to a Wireless Sensor Network (WSN) dataset for subversion of the IDS. Additionally, we explore possible mitigations for the proposed evasion attack, through adversarial example training, outlier detection, and a combination of the two. By using the combination, we are able to reduce the possible attack space by nearly two orders of magnitude.

Keywords: Intrusion detection systems · Adversarial machine learning · IoT · On-demand routing

1 Introduction

As the Internet of Things (IoT) further develops into an all encompassing Internet of Everything, the concept of intelligent networking gains more presence in the field. Intelligent networking is the method of leveraging the data-rich environments highly connected IoT devices create in order to make decisions about topology management and data dissemination paths. Reactive routing is well suited for highly dynamic networks because of the high degree of agility it

© Springer Nature Switzerland AG 2022
E. Renault et al. (Eds.): MLN 2021, LNCS 13175, pp. 85–97, 2022.
https://doi.org/10.1007/978-3-030-98978-1_6

affords; yet reactive routing comes with an increased susceptibility to attacks. The overarching goal of routing layer attacks is to disrupt the on-demand establishment of data dissemination paths by abusing the routing request algorithm so that data packets are forwarded to malicious nodes and dropped before reaching the destination, or replaced with ones that carry false data [11].

Intrusion Detection Systems (IDS) are used to detect attacks targeting routing algorithms. There are three broad types of IDS for routing attacks: Specification Based, Anomaly Based, and Misuse Based. Specification based IDS lay out a set of rules that users must abide by while operating in the network, deviation from which will flag the node as an intruder. A misuse based IDS utilizes predefined attack signatures to identify intruders. For this reason, it is weak against novel attacks as there are no known signatures. Lastly—the focal point of this paper—anomaly based IDS set a level of behavior that is considered as "normal" for a network. Behavior deviating from such a baseline level is considered "anomalous", and is flagged as an intruder [8]. Previously, determining the baseline level of a network has been challenging, especially when considering the many factors at play in an IoT. Additionally, the baseline must be able to adjust over time to cope with network dynamics, e.g. topology change, application requirement, etc.

Due to the vast number of features and amount of generated data that must be analyzed to make an IDS classification, Deep Learning (DL) approaches have begun to become an attractive choice for baseline determination. A DL-empowered IDS consists of observing nodes ingesting packet traffic from the network and classifying this data through a pre-trained DL model [12]. DL-based IDS have been seen as extremely efficient against attackers, boasting accuracy rates of above 99%. However, DL models are vulnerable to Adversarial Machine Learning (AML), which an adversary can employ to attack the network and mislead the IDS classifier. AML attacks take many forms, but the unifying factor behind them is an adversary carefully manipulating the behavior or features used by the model so that wrong classifications are made [9]. In the context of IDS, this goal would be inducing false classification as benign. AML attacks rely upon the adversary's knowledge of the used model.

By knowing the design principle and analysis strategy of the IDS, an attacker would strive to evade them. Particularly, the attacker will seek to introduce selected perturbations to the data collected by the IDS in order to diminish a model's accuracy, and consequently the fidelity of the IDS classification of malicious and benign patterns. Such an attempt is often referred to in the realm of machine learning research as the evasion attack [18]. If an attacker has access to the model used as the gold standard for an IDS, they will be able to see how features are classified by the IDS, and adjust their behavior, and thus their seen features accordingly. Behavior adjustment can be induced by directly modifying controllable features such that the IDS would classify them as benign. Additionally, with knowledge of the model, an attacker can simulate the classification output for different values of directly controllable features for a reactive packet in the system in the case that the original action would have brought on suspicion. An example is the Received Signal Strength Indicator (RSSI) observed by the IDS for an attacking node. If an attacker knows that RSSI is one of the features utilized by the IDS model, they can simulate what different RSSI values will change the classification, and then directly modify their transmission power to adjust their RSSI to match the evading value. While adversarial attacks exist against DL models, various measures also exist to mitigate the possible damage. These measures vary from augmentations to the training dataset through adding adversarial examples, to creating completely separate models in the form of feature squeezing or outlier detection [19, 20]. Many of these defense measures can be easily incorporated in an existing IDS.

This paper focuses on the threat of AML to IDS and introduces a novel intelligent routing layer attack. Particularly, the paper presents an evasion technique that defeats the DL model for an IDS. AML is employed to mislead the detection system by modifying the features the attacker's node directly and indirectly exhibits, and by also deciding when to trigger an attack based on how the IDS operates. To the best of the authors' knowledge this paper contributes the first evasion attack on reactive routing in IoT and demonstrates what different mitigation strategies can be applied and how they can impact the attack success. We experiment with two separate methods to mitigate the attack strength, and combine them to create a strong method to reduce the impact of an adversary in the network. When applied to the IDS, our combined defense reduced the attack success rate by 99%.

The paper is organized as follows. The next section covers related work in the literature. The proposed evasion attack strategy is described in Sect. 3. Sect. 5 presents the validation results. Finally, the paper is concluded in Sect. 6.

2   Related Work

This section highlights both the prior art in the field of IDS for the prevention of routing layer attacks, and adversarial attacks on DL based IDS.

2.1   Deep Learning IDS

The applicability of DL in IDS for wireless sensor networks (WSNs) has been studied in [13]. A dataset generated using ns-3 simulations is used to demonstrate the effectiveness of DL-based IDS and the pros and cons relative to the conventional machine learning (ML) models. The WSN-DS dataset [1] was created to provide a standardized evaluation data for research on routing level attacks in WSNs. The popular LEACH clustering protocol is implemented in ns-3; attacks were then launched and the features exhibited by nodes were tracked during simulation. The findings from routing attack research on LEACH protocols are applicable to other routing protocols specifically created for IoT, like Ad hoc On-demand Distance Vector routing (AODV) or Dynamic Source Routing (DSR). The dataset was created with four different attacks, all of which concentrate on causing Denial of Service (DoS) to a network. These attacks are: (1) Flooding,

(2) Scheduling, (3) Blackhole, (4) Grayhole attacks The authors further implemented a DL-based IDS to demonstrate the utility of the dataset; this IDS will be used as the basis for our model IDS.

Premkumar et al. have developed DLDM, a Deep Learning based Denial of Service (DoS) IDS in [15]. DLDM focuses on seven different types of DoS attacks A number of those highlighted are considered in our work. Specifically, the authors point out that various features related to signal strength and presence in the network are likely to be greatly anomalous for attackers. This would be seen in features like RSSI and packets sent per second by a node. Additionally they indicate that forwarding and packet drop statistics are helpful for detecting attackers.

2.2   Evasion Attack

Evasion is a type of adversarial machine learning methodology that has gained notoriety in recent years [16]. Jiang et al. have studied such an attack against autonomous vehicles in [6], demonstrating how the supply chain of a DL model can be compromised. They apply evasion to the models used to process traffic signs while traveling on the road. The attack is carried out by adding perturbation values to the original data, and monitoring the output of the classifier. The perturbation value with the highest probability of causing misclassification is selected through a Particle Swarm Optimization method. This attack reduces the accuracy of the classifier by a degree of four, demonstrating the grave danger of an evasion attack. While this paper highlights the fact that self-driving cars could be induced to ignore street signs, it is applicable to networking IDS as well. Network IDS operate off of non-uniform feature sets, and have models that are vulnerable to this type of attack. For that reason, it is critical to find potential defenses to reduce the attack space and impact.

Autonomous vehicles are not the only field in which this type of attack can be dangerous. Kolosnjaji et al. present the possibility of malware utilizing evasion in order to skip detection by a DL Model in [7]. In this case, evasion is applied by appending perturbation values to the end of the malware binary. A gradient-descent algorithm is used to determine the possible perturbation for each byte that is added. Each time that a new perturbation value is appended to the binary, it is tested against the test model to monitor if the identification was evaded. Once the optimal perturbation vector is found, it is be appended to the binary. This approach is interesting as it utilizes the threat model of an attacker compromising and stealing the IDS. By doing so, researchers were able to evade detection by an accurate antivirus classifier in 60% of cases. This amount of misclassification is enough to erode trust in the system.

Another successful attack against anti-malware tools is presented in [5], where Huang et al. demonstrate the efficacy of such an attack against both the white-box and gray-box scenarios. In the former, the attackers have full knowledge and access to the system, malware detection model and the utilized features. The gray-box scenario is one in which the attackers only know the features used to train the model, but have no access to the training data or classifying model.

In both scenarios, the attack works in the same fashion, utilizing a Jacobian-based Saliency Map Approach to perturb a number of features monitored by the classifier. By using an optimizer for the adversarial generation, an attacker can reduce the number of features and magnitudes that need to be introduced to produce an anomaly. This can help to prevent being overcome by defense tactics such as feature squeezing, in which unnecessary feature inputs are ignored, and the output with and without those features is compared. In the case that the models' results differ, it can be assumed that an adversarial attack has occurred [19].

2.3   Adversarial Attack Defense

There is also research into how to defend against evasive actors. The play between the adversaries and defenders is noted as an "Arms Race" by [2]. This paper investigates the possibility of retraining of a model to combat attacks, an approach known more generally as Adversarial Example Training. The goal of this is to fold adversarial attack examples into the training sets in order for the model to associate data points resembling an adversarial attack as attacks. They observe that while adversarial example training is sufficient to prevent an attack, the best approach is to increase the cost required to induce a misclassification in the system.

Outlier detection is another powerful tactic to limit the attack space and increase the costs of carrying out an adversarial attack on a model. This is demonstrated in [4], where the authors integrate multiple outlier detection strategies into models to prevent an attack. They use two heuristics to generate data points to classify outliers by: distance of energy costs between differing samples, and the distance between two different features. The rationale from using these two features is that any adversarial attack will be classified as a statistical outlier when compared to known normal samples. We apply the general approach of outlier detection in our research, with this same rationale.

3   Proposed Evasion Attack Strategy

3.1   Oracle Evasion Attack

We propose an adversarial machine learning attack we refer to as the Oracle Evasion Attack (OEA). This attack relies on the premise that the classifying model for the IDS has been either stolen from a compromised node, or inferred through an exploratory attack. This allows the attacker to evaluate how the IDS would classify their behavior against their current features, and perform optimized falsification of those features to fool the IDS. An attacker carrying out the OEA seeks to find a perturbation vector P which, when added to a base truth vector T to produce an evasion vector E, will result in the IDS model producing classification output M with a label of *normal operator*, rather than its true label of *attacker*. We envision that by finding this perturbation vector, the attacker will be able to actively modify their node's behavior and falsify other reported statistics to match the calculated evasion vector. The flow of this attack is illustrated in Fig. 1.

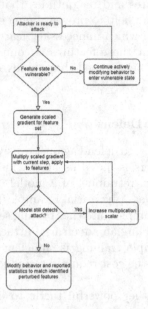

Fig. 1. Flow of Steps for an adversary performing an evasion attack.

Vector P is determined by the following process. First, the features utilized in classification are partitioned into two sets: the set of features directly controllable by nodes DC and those that are indirectly controlled, or uncontrollable UC. A feature $F \in DC$ if and only if a node can directly change its behavior to affect such an observed feature. It is assumed that some features can be controlled by an attacker through falsifying report, e.g., response to a route discovery request. To illustrate, the DC and UC partitions for the WSN-DS dataset [1] are shown in Table 1.

For the purposes of attack detection, some features will inherently be of more importance to the classifier than others. As stated earlier in Sect. 2, the features that indicate the amount of sent data, energy consumption, signal strengths, and number of sent requests are of interest to the classification of routing attackers. The intuition is that routing layer attackers must send out malformed or carefully crafted data packets in order to convince members in the network to route through them. In order to ensure the highest success, the adversary attempts to satisfy two criteria: (i) a high number of routing packets sent, which boosts the statistics for data sent, energy consumption and number of requests, and (ii) a high presence in the network so that others will overhear the data packets routed by the compromised node, thus resulting in a high perceived signal strength and energy consumed. In the aforementioned WSN-DS dataset example, those features are:

Table 1. Feature partitions for WSN-DS dataset. CH is cluster head, and BS is base station.

Uncontrollable (UC) features	Directly Controllable (DC) features
Node ID	RSSI
Time	Distance to CH
Is CH?	$Join_{REQ}$ send
Who is CH?	Data sent
ADV_{CH} send	Data received
ADV_{CH} receives	Data Sent to BS
Rank	Current Energy
Distance between CH,BS	$Join_{REQ}$ recv
Send Code	ADV_{SCH} receives
Max Distance to CH	Energy Consumption
Avg Distance to CH	
ADV_{SCH} send	

1. *RSSI* - a controllable feature that could boosts attacker's network presence.
2. *Data sent* - a feature that can be controlled through traffic shaping; high network presence will cause higher amounts of data to be transmitted.
3. *Data received* - a controllable feature through traffic shaping; a routing attack that requires traffic to flow to the adversarial node, the corresponding received data feature will be anomalously high.
4. *Energy consumption* - To have more network presence, an attacking node consumes high energy in transmitting and receiving packets. This feature is controllable through throttling network presence attempts.
5. *$Join_{REQ}$ send* - A directly controllable feature that is high for nodes attempting to carry out routing attacks.

With the partitioned feature sets, the attacker can determine which features can practically be manipulated or misreported with a calculated perturbation vector. To elaborate, when calculating P, values that cannot be controlled by the node may be affected. This causes a misclassification against the model, yet it is impossible to implement. For this reason, P shall have a perturbation value of zeros for features belonging to UC and a non-zero value for those of DC.

3.2   Modified FGSM

Considering DC, the attacker could apply a modified version of the Fast Gradient Sign Method (FGSM), an AML procedure first proposed by Goodfellow et al. [3]. The idea of FGSM is that by adding a perturbation vector whose entries correspond to the sign of the gradient vector of the cost function multiplied by a scalar, one will be able to disturb classification of an input, thus allowing an attacker to evade detection and stealthily disrupt the network operation. Given x as the input to the model, y as the classes for the model (in our case will be malicious and benign) and θ as the parameters for the model, the perturbation vector will be calculated as: $sign(\nabla_x J(\theta, x, y)) \cdot \epsilon$, where $J(\theta, x, y)$ is the cost function to train the network and ϵ is the magnitude by which to perturb the input data. FGSM is highly effective for classifications in which the features are all similarly measured, e.g., image classification based on pixels. However, for data classification where different feature measurements can vary by orders of magnitude, different challenges are faced when applying FGSM. The main challenges for applying FGSM to IDS datasets are:

1. The magnitude of features is not constant throughout the set. Hence, the ϵ scalar value applied to the perturbation vector will cause a greatly imbalanced difference between the original input and perturbed input. For this reason, each feature will need its own specific ϵ.
2. The feature data types may differ. Perturbation of a Boolean feature is limited to a value of 0 or 1; in other data types with limited ranges large perturbation could induce invalid feature values.
3. Some features are immutable by the attacking node. This significantly constrains FGSM, where the ϵ for features that are directly controllable by an attacking node must be increased in order to compensate for features that are uncontrollable, which will have ϵ of 0.

The attacker begins with the gradient vector calculated from the FGSM, $sign(\nabla_x J(\theta, x, y))$ and then calculates a perturbation ϵ vector by the following. First, the attacker compiles a sample set of data points for the features S_S, and calculates the arithmetic mean for each feature, creating a vector to represent them $\overline{S_S}$. The $\log_{10}$ is then taken for each element of $\overline{S_S}$, For each element in $\log_{10} \overline{S_S}$, if the feature belongs to the set UC, the element is set to zero, as that feature cannot be perturbed. An adversary will apply the following steps to carry out OEA:

1. The attacker generates a gradient vector $sign(\nabla_x J(\theta, x, y))$ using FGSM procedure for each test point, and calculates the bounding vector B_P.
2. The gradient vector is multiplied by a binary vector that sets perturbation to 0 for every feature in the UC set.
3. Starting with a perturbation factor of 1, the attacker then steps by one through a set factor limit. This limit is set to cut off the generation process when a data point cannot be perturbed to induce a different classification within a feasible amount.
4. For each step between a perturbation factor of 1 and the limit, the attacker tests the perturbed data point with the IDS model. When the IDS misclassifies the sample, the attacker stops stepping through factors.

The attacker now knows that if the data feature they are exhibiting is perturbed by the gradient vector multiplied by the found scalar perturbation factor, the IDS in the system under attack will classify the malicious node as benign, and they shall avoid detection.

4 Defense Strategy

In order for an IDS to mitigate the threat of an evasion attack generated through adversarial tactics, it must be augmented with defensive strategies. In this work we study adversarial example training and outlier detection, when applied individually and in combination.

4.1 Adversarial Example Training

This strategy is based on training the model using perturbed samples, and is proven to be quite effective [10]. Augmenting the training pipeline with a perturbed dataset fills the gaps between the decision sets, preventing an attacker from employing FGSM to misclassify a sample. While utilizing the output of the FGSM for adversarial samples to perform training is considered by many as being too weak, Wong et al. [17] find that simply changing certain input seeds and transformations is enough to generate a strong defense training set. For that reason, we believe that generating perturbations using our Modified FGSM provides sufficient protection, due to the variable scalar and differently weighted features in the gradient.

4.2 Outlier Detection

This strategy is often used when attackers are able to manipulate their features that are utilized for an input to the IDS. By applying OEA, an attacker is able to generate samples with uncharacteristically high perturbation scalars that would cause the desired misclassification. It would be ideal for a system to detect such perturbed inputs as outright examples of an attacker, due to their outlier behavior. This strategy is usually implemented as a separate model that would be run on inputs, with any outlier being detected as an attack attempt. To create the outlier detection model for the considered WSN-DS dataset, we used the three-layer densely connected neural network previously mentioned. However, to train this network, we create a new dataset consisting of the original WSN-DS data, and corresponding adversarial examples for that dataset to be considered as outliers. The data is labeled by whether it is legitimate data, or a generated example. The model is then trained for binary classification on this training data, with 0 corresponding to a benign sample, and 1 corresponding to an adversarial outlier.

5 Validation Results

To validate the impact of OEA and the efficacy of the defense strategies, the attack was implemented utilizing the Cleverhans python library [14] to aid in generation of gradient vectors. A deep learning IDS model was created with the WSN-DS dataset. This model configuration was chosen to keep in line with other IDS models based on the dataset, mainly the model proposed in the WSN-DS

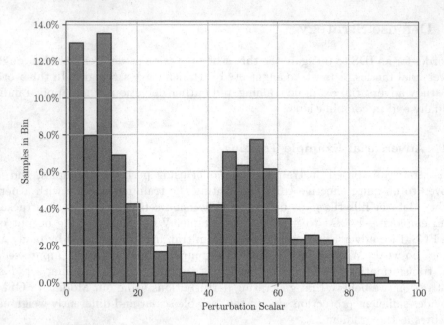

Fig. 2. Histogram of perturbation scalar in attack scenario

paper [1], which consists of three densely connected neural networks. This model is highly effective at accurately detecting multiple types of routing attacks, with an overall accuracy of 97.5431% when tested against the WSN-DS. The dataset was split to have two-thirds of the set as the training set, and one-third as the validation set. Training yielded a model that had a base accuracy of 97.31% with a loss of 0.058. Training was carried out over 10 epochs, and loss was calculated using the categorical cross-entropy method.

The modified FGSM technique was applied against the trained model; the scalar that the gradient power vector is multiplied with has been incrementally grown from 1 through 250. A possible attacking perturbation was found in 57,504 out of 123,638 samples. This accounts for 46.51% of samples in the test subset. While this limits the potential effect on the overall accuracy of the IDS model, it is a large enough percentage of samples that can be perturbed in an attack. The skew of the scalar necessary to be applied to the gradient to induce a change in classification is represented in the histogram in Fig. 2. When generating this histogram, scalars of zero were excluded, as they simply constitute examples of where the model is inherently inaccurate, and not a result of an attack. Observing the histogram, we see a heavy skew towards smaller values, suggesting that most feature states that are vulnerable to perturbation do not need to have a large scalar applied to them. Hence, it can be concluded that with an unprepared model, it does not take a large change of exhibited features to mislead the classification model.

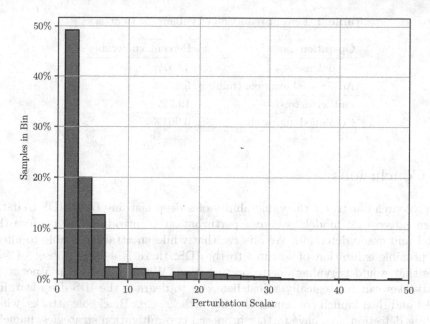

Fig. 3. Histogram of perturbation scalar in combined defense scenario

The results of the various adversarial attack defenses are recorded in Table 2. When performing our adversarial training defense, we observe that the percentage of samples open to perturbation is reduced from 46.51% to 6.857%. Additionally, the distribution of required scalar perturbation is spread more evenly between the minimum and maximum set scalar values. This distribution is demonstrated in the histogram in Fig. 3. Meanwhile, by applying our outlier detection method, we were able to accurately detect attacks with an 80.73% accuracy rate. This model would ideally be run in conjunction with a conventional IDS to flag whether any data point could be modified for the purpose of misleading the IDS. Combining the two defense mechanisms gives a total reduction of samples open to perturbation from 46.51% to 0.961%. This is a satisfactory reduction and certainly effective for mitigating the attack threat to the system. Consider the idea that the vulnerable data points are states that an attacker would recognize that they are close to. When within a certain distance of exhibiting these features, they can then actively modify their behavior to meet being similar enough to carry out an evasion attack. When only 0.961% of test data points are vulnerable to this style of attack, it severely reduces the possible attack space for an adversary to the point where the cost to carry it out on a network is exorbitant.

Table 2. Percent reduction of vulnerable to attacks

Operation mode	Percent vulnerable
No defense	46.51%
Adversarial example training	6.857%
Outlier detection	19.72%
Combined approach	0.961%

6   Conclusions

This research illustrates the vulnerability of a deep learning based IDS to data-driven adversarial models, where a perturbation is introduced to confuse the model and evade detection. We observe that while an attacker is able to affect any possible exhibition of features to the IDS, there is a large subset of features that would be vulnerable to perturbation. When operating in a network, an attacker can strategically adjust behavior to degrade the IDS deep learning model and then launch conventional cyber attacks, e.g., Blackhole attacks, while evading detection. We have further proposed two mitigation strategies, namely, adversarial example training and outlier detection. Using a publicly available dataset, we have demonstrated the effectiveness of our defense strategies where the potential attack space is diminished to a point where the proposed attack model would be infeasible.

References

1. Almomani, I., Al-Kasasbeh, B., Al-Akhras, M.: Wsn-ds: a dataset for intrusion detection systems in wireless sensor networks. J. Sens. **2016** (2016)
2. Chen, L., Ye, Y., Bourlai, T.: Adversarial machine learning in malware detection: arms race between evasion attack and defense. In: 2017 European Intelligence and Security Informatics Conference (EISIC), pp. 99–106. IEEE (2017)
3. Goodfellow, I.J., Shlens, J., Szegedy, C.: Explaining and harnessing adversarial examples. arXiv preprint arXiv:1412.6572 (2014)
4. Grosse, K., Manoharan, P., Papernot, N., Backes, M., McDaniel, P.: On the (statistical) detection of adversarial examples. arXiv preprint arXiv:1702.06280 (2017)
5. Huang, Y., Verma, U., Fralick, C., Infantec-Lopez, G., Kumar, B., Woodward, C.: Malware evasion attack and defense. In: 2019 49th Annual IEEE/IFIP International Conference on Dependable Systems and Networks Workshops (DSN-W), pp. 34–38. IEEE (2019)
6. Jiang, W., Li, H., Liu, S., Luo, X., Lu, R.: Poisoning and evasion attacks against deep learning algorithms in autonomous vehicles. IEEE Trans. Veh. Technol. **69**(4), 4439–4449 (2020)
7. Kolosnjaji, B., et al.: Adversarial malware binaries: evading deep learning for malware detection in executables. In: 2018 26th European signal processing conference (EUSIPCO), pp. 533–537. IEEE (2018)

8. Korba, A.A., Nafaa, M., Salim, G.: Survey of routing attacks and countermeasures in mobile ad hoc networks. In: 2013 UKSim 15th Int'l Conference on Computer Modelling & Simulation, pp. 693–698. IEEE (2013)
9. Kurakin, A., Goodfellow, I., Bengio, S.: Adversarial machine learning at scale. arXiv preprint arXiv:1611.01236 (2016)
10. Madry, A., Makelov, A., Schmidt, L., Tsipras, D., Vladu, A.: Towards deep learning models resistant to adversarial attacks. arXiv preprint arXiv:1706.06083 (2017)
11. Nadeem, A., Howarth, M.P.: A survey of manet intrusion detection & prevention approaches for network layer attacks. IEEE Commun. Surv. Tutor. **15**(4), 2027–2045 (2013)
12. Nguyen, S.N., Nguyen, V.Q., Choi, J., Kim, K.: Design and implementation of intrusion detection system using convolutional neural network for dos detection. In: Proceedings of the 2nd International Conference on Machine Learning and Soft Computing, pp. 34–38 (2018)
13. Otoum, S., Kantarci, B., Mouftah, H.T.: On the feasibility of deep learning in sensor network intrusion detection. IEEE Netw. Lett. **1**(2), 68–71 (2019)
14. Papernot, N., et al.: Technical report on the cleverhans v2.1.0 adversarial examples library. arXiv preprint arXiv:1610.00768 (2018)
15. Premkumar, M., Sundararajan, T.: Dldm: deep learning-based defense mechanism for denial of service attacks in wireless sensor networks. Microprocess. Microsyst. **79**, 103278 (2020)
16. Shi, Y., Sagduyu, Y.E.: Evasion and causative attacks with adversarial deep learning. In: MILCOM 2017–2017 IEEE Military Communications Conference (MILCOM), pp. 243–248. IEEE (2017)
17. Wong, E., Rice, L., Kolter, J.Z.: Fast is better than free: Revisiting adversarial training. arXiv preprint arXiv:2001.03994 (2020)
18. Xu, G., Li, H., Ren, H., Yang, K., Deng, R.H.: Data security issues in deep learning: attacks, countermeasures, and opportunities. IEEE Commun. Mag. **57**(11), 116–122 (2019)
19. Xu, W., Evans, D., Qi, Y.: Feature squeezing: detecting adversarial examples in deep neural networks. arXiv preprint arXiv:1704.01155 (2017)
20. Yuan, X., He, P., Zhu, Q., Li, X.: Adversarial examples: attacks and defenses for deep learning. IEEE Trans. Neural Netw. Learn. Syst. **30**(9), 2805–2824 (2019)

DynamicDeepFlow: An Approach for Identifying Changes in Network Traffic Flow Using Unsupervised Clustering

Sheng Shen [1,2P(✉)], Mariam Kiran[1], and Bashir Mohammed[1]

[1] Lawrence Berkeley National Laboratory, Berkeley, CA 94720, USA
mkiran@es.net, bmohammed@lbl.gov
[2] Tula Technology, San Jose, CA 9513, USA
shens@tulatech.com

Abstract. Understanding flow changes in network traffic has great importance in designing and building robust networking infrastructure. Recent efforts from industry and academia have led to the development of monitoring tools that are capable of collecting real-time flow data, predicting future traffic patterns, and mirroring packet headers. These monitoring tools, however, require offline analysis of the data to understand the big versus small flows and recognize congestion hot spots in the network, which is still an unfilled gap in research. In this study, we proposed an innovative unsupervised clustering approach, DynamicDeepFlow, for network traffic pattern clustering. The DynamicDeepFlow can recognize unseen network traffic patterns based on the analysis of the rapid flow changes from the historical data. The proposed method consists of a deep learning model, variational autoencoder, and a shallow learning model, k-means++. The variational autoencoder is used to compress and extract the most useful features from the flow inputs. The compressed and extracted features then serve as input-output pairs to k-means++. The k-means++ explores the structure hidden in these features and then uses them to cluster the network traffic patterns. To the best of our knowledge, this is one of the first attempts to apply a real-time network clustering approach to monitor network operations. The real-world network flow data from Energy Sciences Network (a network serving the U.S. Department of Energy to support U.S. scientific research) was utilized to verify the performance of the proposed approach in network traffic pattern clustering. The verification results show that the proposed method is able to distinguish anomalous network traffic patterns from normal patterns, and thereby trigger an anomaly flag.

Keywords: Network traffic pattern clustering · Unsupervised learning · Flow recognition · Wide area network

1 Introduction

Wide area network (WAN) service providers are dealing with relentless traffic growth, making it a challenge to manage capacity, recognize anomalies, and

© Springer Nature Switzerland AG 2022
E. Renault et al. (Eds.): MLN 2021, LNCS 13175, pp. 98–116, 2022.
https://doi.org/10.1007/978-3-030-98978-1_7

balance performance with the dynamic network traffic patterns [1]. Especially during the COVID era, the proliferation of video streaming and mobile applications, as well as an increase in work-from-home practices have introduced new data sources and destinations to the network. Thus, it is essential to understand the spatial and temporal distributions of the network flows, which can provide an early indication of sudden surges in network traffic and an opportunity to implement measures to prepare for these unseen network traffic patterns [2].

Current network monitoring services are designed to monitor the amount of data (gigabytes) transmitted, the number of packets, or the bandwidth they occupy. Network traffic monitoring models that simulate current and predict future traffic patterns have been useful for solving flow capacity planning challenges [3,4]. For example, the use of deep learning models to predict network traffic patterns seven days in advance can assist engineers in identifying congestion points and avoiding them [5]. Congestion, in particular, can result in packet loss and impaired performance in large scientific transfers [6]. In addition, the recent development of industry automation and connected devices has created a demand for massive amounts of network resources, resulting in an increase in flow exchange and new patterns of network traffic [7]. Therefore, it is imperative to have a network traffic monitoring model that can analyze flow change behavior and identify new network traffic patterns, enabling engineers to be aware of the upcoming network changes.

In this study, we developed a novel DynamicDeepFlow Neural network (DDF) approach for network traffic pattern clustering. The proposed method can extract useful features from spatial and temporal network traffic flow data, and also raise anomalies for any changes that occur in the network traffic pattern. The DDF consists of a deep learning model, variational autoencoder (VAE), and a shallow learning model, k-means++. Specifically, this work advances the state-of-the-art in network traffic clustering for WAN research as,

- We propose a novel DDF approach for anomalous network traffic pattern clustering. The proposed method takes advantage of both the neural network and unsupervised clustering method.
- Using Pearson correlation, we capture the spatial and temporal network traffic flow change information among all the sites in WAN.
- By using an VAE model, this study automates the feature learning process from a large amount of network flow data. The learned features enable k-means++ to cluster anomalous network traffic patterns.

Our results indicate that the proposed method has good generalization capability in clustering the network traffic patterns. This capability is to produce reasonable outputs for new network traffic patterns that have never been seen during the training process.

2   Related Work

Clustering of network traffic patterns is essential for network management, as it discloses the source, destination, and other factors such as quality of service, network security, traffic visualization, etc. Recent studies [8] have reported many interesting approaches for network traffic pattern identification, such as port-based, payload-based, and flow statistics-based. Port-based methods use port information for service identification but are unreliable since internet services may not operate on well-known ports. To solve this problem, the payload-based methods search well-known patterns inside the packets by employing deep packet inspection. However, the increased cost of maintaining the up-to-date pattern database is a challenge to payload-based methods. Flow statistics-based methods rely on high-level information of the packet header and thus make a good choice to approach the downsides of non-available payloads and dynamic ports. This paper mainly focuses on the flow statistics-based methods. Most of them can be categorized into supervised learning methods and unsupervised learning methods.

Supervised learning methods build network pattern classification models based on data that is labeled by network experts. These models classify unseen data to a class either by directly learning the boundary between classes or by learning the distribution of data in a class using Bayes rules. Soule et al. [9] performed flow classification to identify network pattern among a small set of classes (long-life and large data flow or smaller versions). Bayesian methods use the Bayes rule as a starting point to generate a posterior class distribution based on a prior class distribution and the observed attributes. Auld et al. [10] proposed a Bayesian neural network in the field of network traffic classification without access to the contents of packets. Michael et al. [5] proposed an artificial neural network for network traffic classification which is capable of achieving high classification accuracy and maintaining low system throughput. A key advantage of the neural network over traditional machine learning schemes is the introduction of non-linearity which can better capture complexities within class structures. However, supervised learning methods require a large amount of labeled data derived from network human expertise that is sometimes unavailable and costly.

Unsupervised learning methods perform network classification by analyzing and clustering unlabeled data. These methods discover the patterns hidden in data without the need for human intervention. Munz et al. [11] proposed a novel flow-based network identification scheme based on the K-means clustering algorithm. The proposed k-means algorithm was able to cluster normal and anomalous network traffic patterns. Ahmed et al. [12] proposed a framework for anomalous network traffic patterns detection using a novel clustering technique. However, they failed to address the issue of detecting the anomaly under dynamic network traffic.

Although these approaches produce satisfactory results in clustering network traffic pattern, they do not address the problem of detecting changes in flows within a dynamic network environment. As a result of the fast-changing dynamic nature and non-linearity of network patterns, it is challenging to model the network pattern using these traditional network traffic clustering approaches. This study thus presents a novel DDF approach for network traffic pattern clustering.

The proposed model is capable of online identifying anomalous network traffic patterns in a dynamic, rapidly changing network environment.

3 Motivation

The WAN is a large computer network that connects individuals or groups of computers over a wide geographical area for the purpose of sharing information. The Energy Sciences Network (ESnet) is a WAN facility, which allows scientists to transfer their experimental results from complex experimental facilities to their collaborators throughout the world. However, issues such as packet loss and delay in the network may jeopardize their experimental results. Accordingly, it is essential to receive the transferred data at the appropriate time for further processing and maintenance of the network infrastructure in order to achieve optimal network performance.

We modeled the WAN as a graph $G = (V, E, A)$, where V is a set of n nodes that represents sites; E is a set of directed edges representing the connection between nodes, and $A \in R^{N \times N}$ is the weighted adjacency matrix representing the strength of connectivity between nodes. The weight adjacency matrix can represent the spatiotemporal data collected at timestamps, to get a snapshot of the network at the time.

The challenge in network planning is to find dynamic flow changes, which send as much as flow from the source (s) to destination (d) in a given time (T) and if this pattern has changed. In this study, we are not taking into account the flow transit time. But the total number of flows active during a time interval θ. We make this assumption because we only want to learn if common network traffic patterns have changed during certain hours, daytime or nighttime.

Definition 1 (Flows over θ time). A network snapshot during time horizon θ consists of all participating flows entering sources S and leaving all destinations D during θ. Figure 1 shows the snapshots of network flows over time. This study consider each snapshot as an independent network image.

Equation 1 is the Pearson correlation, where $(\sigma_{rk_A}, \sigma_{rk_B})$ are the standard deviations of the rank variables, $cov(rk_A, rk_B)$ is the covariance of the rank variables, and ρ is the pearson correlation coefficient applied to the rank variables. r_s ranges from -1 to 1. 0.5 and above indicate a positive correlation among two traffic traces, while -0.5 indicates a negative correlation.

$$r_s = \rho_{rk_A, rk_B} = \frac{cov(rk_A, rk_B)}{\sigma_{rk_A}, \sigma_{rk_B}},\qquad(1)$$

By the Definition 1, we are motivated to cluster new patterns of network traffic in time horizon θ, which have previously not been observed by the network. These unseen network traffic patterns are noted as anomalies, which may be the result of deteriorating network performance or merely a traffic pattern that was previously observed during the day but has now been spotted at night. The ability to cluster these anomalous network traffic patterns enables us to address security concerns, perform new data movement that was not anticipated, or optimize future capacity planning.

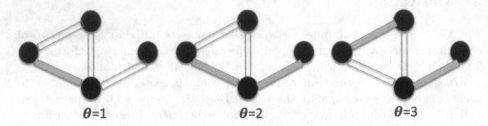

Fig. 1. Snapshots of network flows on a topology over time horizon $\theta = \{1,2,3\}$. The shading shows the occupied links.

Figure 2 illustrates the pipeline for building a machine learning for network traffic pattern clustering. The pipeline consists of two steps, offline training and online inference. Specifically, we pre-trained a machine learning model with the flow change correlation in years 2018 and 2019. The pre-trained machine learning model was then used to cluster the new network traffic patterns that may occur in year 2020. We assume that 2020 may happen some unique network traffic patterns as fewer employees are being physically in the office due to the work from home policies.

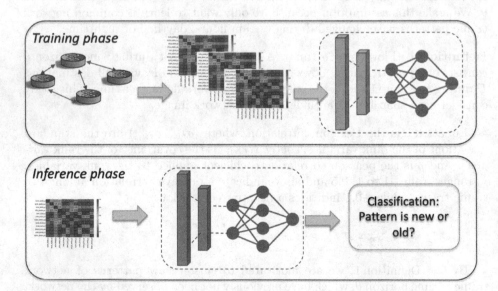

Fig. 2. The machine learning pipeline for flow pattern recognition.

4   Building DynamicDeepFlow Neural Network

4.1   Input and Output Structures

The flow data was measured based on 52 research sites and universities in two-way traffic traces as they use separate fiber-optic links. A total of 104 traffic flow data was thus collected and each refers to the amount of incoming or outgoing flow between two research sites and universities. Pearson correlation coefficient was employed to derive the network traffic flow relationships among the 52 research sites and universities [13]. As such, the input to the model was a Pearson correlation coefficient matrix with fixed size 104 × 104, of which the rows and columns were associated to the 104 incoming and outgoing flow data. The matrix is defined as

$$Input = \begin{bmatrix} r_{1,1} & r_{1,j} & \cdots & r_{1,104} \\ r_{i,1} & r_{i,j} & \cdots & r_{i,104} \\ \vdots & \vdots & \ddots & \vdots \\ r_{104,1} & r_{104,j} & \cdots & r_{104,104} \end{bmatrix} \tag{2}$$

Figure 3 shows a heat map of the Pearson correlation coefficient between 12 research sites. The level of linearity between two sites is expressed using color and coefficient value. The stronger the linear relationship, the redder the color and the closer the value to 1. Conversely, the closer the value is to 0 and the bluer the color, the weaker the linear relationship. For example, HOUS_NASH_out has a high correlation with ELPA_SUNN_in with a value of 0.94. In another case, ELPA_SUNN_in has a weak relationship with EASH_WASH_in with a value of 0.13. The _in and _out refer to incoming and outgoing flows.

The VAE model took this matrix as an input and then aimed to reconstruct a similar input using the extracted features during the training process. This process was realized by comparing the differences between the reconstructed inputs and the original inputs. To this end, each matrix served as both the input and true output of the VAE model, described in the next subsection.

4.2   Overall Architecture of DDF

The DDF approach consisted of a deep learning model, VAE and a shallow learning model, k-means++. The VAE investigated the useful features of the network flow and compresses them into low dimensional vectors that can be easily analyzed by the shallow learning model. The k-means++ then was employed to explore the complex dependency of the network traffic pattern on these compressed features extracted from the network flow changes between connected research sites. The overall structure of the DDF is shown in Fig. 4.

An input matrix $\mathbf{X}$ was first fed into the VAE model. The convolutional layer filtered $\mathbf{X}$ with k kernels $\mathbf{K}$ (also called filters) of size $k_h \times k_w \times k_w$ with a stride of size $s_h \times s_w$ via the convolution operation $C(\mathbf{X}, \mathbf{K})$ [14]. The filter moved along $\mathbf{X}$ horizontally and vertically, repeating the same computation for

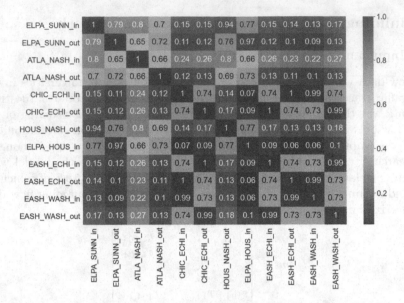

Fig. 3. Pearson correlation coefficient matrix of 12 research sites and universities.

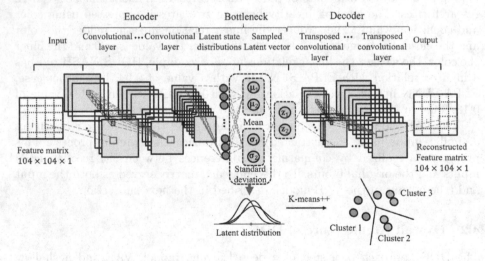

Fig. 4. DDF model architecture. Showing the autoencoder model.

each region, namely, convolving $\mathbf{X}$. The ith row and j^{th} column of kth output of convolutional layer l^{conv},

$$Z_{i,j,k}^{l_{conv}} = C(\mathbf{X}, \mathbf{K})_{i,j,k} \sum_{r=1}^{k_h} \sum_{s=1}^{k_w} \sum_{t'=1}^{k_c} x_{i^{i},j^{i},t^{i}} k_{r,s,t^{i},k} + b_k \qquad (3)$$

$$i' = (i-1).s_h + r \qquad (4)$$

$$j' = (j-1).s_w + s \tag{5}$$

Where $k_{r,s,t^i,k}$ and b_k were the weights and bias of the kth kernel in the convolutional layer, respectively.

After passing through a stack of convolutional layers, the input matrix $\mathbf{X}$ reached the bottleneck. The bottleneck used the low-dimensional latent vector to represent the high-dimensional matrix $\mathbf{X}$. The posterior of the latent vectors is approximated by a multivariate Gaussian that can be expressed as

$$\log q_\phi(\mathbf{z}|\mathbf{x^{(i)}}) = \log N(\mathbf{z}; \mu^{(i)}, \sigma^{2(i)}\mathbf{I}) \tag{6}$$

Where $\mu^{(i)}$ and $\sigma^{(i)}$ were the posterior mean and standard deviation, respectively, $\mathbf{x}^{(i)}$ indicates the ith input matrix. The latent vectors were sampled from the posterior $\mathbf{z}^{(i,l)} \sim q_\phi(\mathbf{z}|\mathbf{x}^{(i)})$ using $\mathbf{z}^{(i,l)} = g_\phi\left(x^{(i)}, \epsilon^{(i)}\right) = \mu^{(i)} + \sigma^{(i)} \odot \epsilon^{(l)}$ where $\epsilon^{(i)} \sim N\left(\mathbf{0}, \mathbf{I}\right)$. $\odot$ indicates an element-wise product. Both $p_\phi(z)$ (the prior) and $q_\phi(z|x)$ followed the Gaussian distribution. To this end, the loss function of the VAE was expressed as

$$L(\theta, \phi; \mathbf{x}^{(i)}) \simeq \frac{1}{2}\sum_{j=1}^{J}\left(1 + \log((\sigma_j^{(i)})^2 - (\mu_j^{(i)})^2 - (\sigma_j^{(i)})^2\right) + \frac{1}{L}\sum_{l=1}^{L}\log p_\theta(\mathbf{x}^{(i)}|\mathbf{z}^{(i,l)})$$

$$\tag{7}$$

where $\mathbf{z}^{(i,l)} = \mu^i + \sigma^i \odot \epsilon^l$ and $\epsilon^i \sim N(0, \mathbf{I})$.

Variational Autoencoder. Autoencoder (AE) is an unsupervised neural network where the inputs have the same size as the outputs. It compresses the input to a single lower-dimensional representation and then reconstructs the inputs using this compressed representation. An AE consists of three components: encoder, latent vector (i.e., bottleneck), and decoder. The encoder compresses the input and puts the compressed inputs to the bottleneck. The latent vector leverages fewer neurons to represent the most important features of the inputs, compressing features. The decoder, which follows the latent vector, reconstructs the inputs using these compressed features. VAE is a variant of AE, which represents the input as a distribution (e.g., Gaussian distribution) instead of a single representation by AE. This distribution enables VAE a capability to generate new samples based on the parameters of the distribution. To this end, the main purpose of the VAE is to identify the optimal parameters of the distribution (e.g., mean and standard deviation). In this study, we assume the variation of traffic flow data approximates a multivariate Gaussian distribution. The structure of the VAE used in this study is shown in Fig. 4. The VAE mainly consists of three types of layers: the convolutional layers, the transposed convolutional layers, and the fully-connected layers.

The convolutional layers are used to execute a special kind of linear operation named convolution [15]. Convolution is an element-wise multiplication between the inputs and filters. Specifically, each unit of a convolutional layer is connected to local patches in the feature maps of the previous layer through a set of weights called filter banks. The result of this locally weighted sum is then passed through a variety of layers, such as a rectified linear unit (ReLU), to form the feature maps of the next layer. The inputs go across multiple convolutional layers and reach two parallel fully-connected layers, which represent the mean and standard deviation of the multivariate Gaussian distribution, respectively. The fully connected layer performs matrix multiplication with separate weight matrices describing the pair-wise interactions between all the input and mean and standard deviation units. The latent vector is thus identified using the learned mean and standard deviation. The latent vector is a highly compacted representation of the input and it can provide variation to the model by sampling from the multivariate Gaussian distribution (i.e., prior distribution). Finally, the decoder reconstructs the input data given the low-dimensional latent representation. This process performs using a stack of transposed convolutional layers. The transposed convolutional layers execute a reverse of the convolution operation, where each unit of the input is connected to the feature maps of the next layer through filters of the transposed convolutional layer. Essentially, it is an upsampling operation by the convolution.

k-means++: The k-means method is a widely used clustering technique that aims to identify k cluster centers by iteratively minimizing the average squared distance between data points and the k cluster centers [16,17]. Each point is then assigned to the nearest cluster center and this process is repeated until newly cluster center keeps constant or the maximum number of iterations are reached. It is of great importance to properly initialize the cluster center. The k-means++ is a variant of k-means that allows adaptive selection of cluster centers based upon their contribution to the cost L.

In this study, the learned features (mean and standard deviation) from the VAE are fed into to the k-means++. The k-means++ is used to explore the structure hidden in these features and determine k clusters for normal and anomalous network traffic patterns. The optimal number of k are experimentally investigated using an elbow method.

4.3   Training Algorithm

The VAE contains a set of unknown parameters (e.g., weights and bias) that need to be optimized and identified in the model training process. To properly identify these parameters, a cost function L was defined to measure differences (or generalization errors) between the model predictions and the associated ground truth.

The expected generalization error given by the cost function was minimized by a widely used optimization method, named adaptive moment estimation (Adam), which is an adaptive learning rate estimation method. It computes individual learning rates for different parameters. Adam can be looked at as a combination of Root Mean Square Propagation (RMSprop) and Stochastic Gradient Descent (SGD) with momentum. It uses the squared gradients to scale the learning rate like RMSprop and also takes advantage of momentum using the moving average of the gradient instead of the gradient itself like SGD with momentum. Adam updated the parameters θ (weights ω, and biases b), to minimize the generalization error by iterative moving in the opposite direction of the gradient, which is the direction of the steepest descent.

5   Experimental Data and Implementation Details

5.1   Experimental Data

The experimental data used in this study is collected from the ESnet, which is a high-speed network designed to support scientific research worldwide. Figure 5 shows the ESnet topology, divided into five time zones. The ESnet provides services to more than 50 research sites and universities, including the entire national laboratory system, its supercomputing facilities, and its major scientific instruments, enabling geographical-free collaboration worldwide.

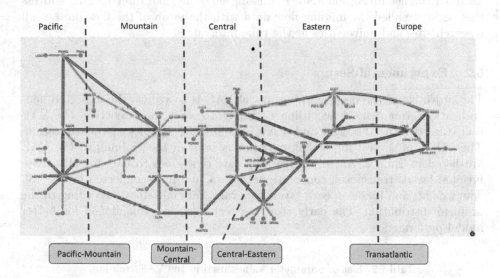

Fig. 5. ESnet network topology (www.es.net).

We monitor link capacity and flow data movement across routers using simple network management protocol (SNMP). Table 1 shows the collected flow data on two-way transfer between Sunnyvale and Sacramento, which are the research sites in the ESnet. The flow data is collected in moving Gigabytes (GBs) across router interfaces in one-hour intervals.

Table 1. Sample timestamped SNMP traffic aggregated from Sacramento and Sunnyvale Interfaces. Note the data was collected in pacific standard time zone (PST).

Timestamp (PST)	SACR_SUNN_in (GB)	SACR_SUNN_out (GB)
01/01/2019 9AM	14,110,930,202	1,025,131,246
01/01/2019 10AM	13,453,619,303	9,191,557,943
01/01/2019 11AM	12,168,879,944	7,793,842,045
01/01/2019 12AM	11,231,198,033	7,097,237,528
01/01/2019 1PM	10,780,847,622	8,048,293,939

The flow data spans from January 1st, 2018 to December 31st, 2020, consisting of 52 research sites and universities, with 104 traffic traces of two-way traffic traces with incoming and outgoing. Due to router interface errors and disconnected Ethernet links, flow data is missing at some time intervals. To address this issue, we filled the missing flow data with the mean of the flow data of all research sites and universities at the time interval.

5.2   Experimental Setup

The objective of the deep learning model VAE is to reduce the expected generalization error defined according to Eq. 7. To meet this objective, the VAE was trained with 30 epochs and a mini-batch of 32 samples at each iteration. The values of several important parameters used in the training of the VAE are listed in Table 2. An initial learning rate α was set to 0.01 for all convolutional layers, transposed convolutional layers, and fully-connected layers [18]. The weights and biases in each layer were randomly initialized according to the uniform distribution. The early stopping technique was abandoned for better model performance.

Table 2. List of parameter values used in the VAE training.

Parameter	Value
Initial learning rate, α	0.01
Minibatch size, γ	32
Beta 1, β_1	0.9
Beta 2, β_2	0.999
epsilon	$1e^{-8}$
L2 Regularization, λ	0.0001
Number of epochs	30

The overall architecture of the VAE model consisted of five convolutional layers, five transposed convolutional layers, and three fully connected layers. Each convolutional layer and transposed convolutional layer were followed by a ReLU except the last transposed convolutional layer, which equipped a sigmoid activation function to ensure the outputs fall in the range [0, 1]. The ReLU was applied to introduce non-linear properties into the neural networks. The first two parallel fully-connected layers directly connected the third fully-connected layer without any activation function and dropout layers. The output of the last transposed convolutional layer provided the reconstruction result corresponding to the original input. The VAE model obtained 926,849 parameters (i.e., weights and bias) and about 8,800 neurons. The configurations of the VAE model are outlined in Table 3.

Table 3. Summary of the configuration of the VAE.

Layername	Filter size	No. of kernel	Size	No. of weights	No. of biases
Input	1 × 104 × 104	–	–	-	–
Conv-1	1 × 5 × 5	32	(3,3)	800	32
Conv-2	1 × 4 × 4	64	(2,2)	32,768	64
Conv-3	1 × 4 × 4	128	(2,2)	131,072	128
Conv-4	1 × 3 × 3	128	(1,1)	147,456	128
Conv-5	1 × 3 × 3	128	(1,1)	147,456	128
FC-1	2 × 1	–	–	2304	0
FC-2	2 × 1	–	–	2304	0
FC-3	1152 × 1		–	2304	0
Transposed conv-1	1 × 3 × 3	128	(1,1)	147,456	128
Transposed conv-2	1 × 3 × 3	128	(1,1)	147,456	128
Transposed conv-3	1 × 4 × 4	64	(2,2)	131,072	64
Transposed conv-4	1 × 4 × 4	32	(2,2)	32,768	32
Transposed conv-5	1 × 5 × 5	1	(3,3)	800	1
Output	1 × 104 × 104	–	–	–	–

The conventional data splitting strategy first shuffles the entire dataset and then splits it into training, validation, and test subsets with a ratio. The data shuffling process enables the training, validation, and test subsets to have a similar data distribution. The successive samples typically are more likely correlated with their neighbors than those arranged farther away [19]. However, the objective of the proposed method is to identify the unseen network traffic pattern, and thus it is necessary to ensure the model does not know the new network traffic pattern before the test process. In other words, the test subset should be made up of complete data from one specific year instead of partial data after shuffling. It is worth mentioning that the current coronavirus pandemic crisis has altered the network traffic patterns, where a surge in incoming network traffic for email, web, and VPN connections is due to the work from the home policy since February 2020. Fewer employees in the office and have to remotely access company infrastructure via a network. To this end, instead of shuffling the entire dataset, we first took the data from 2020 out of the entire dataset, to be used as the test set, and then shuffled the remaining data from 2018 and 2019 to create the training (85%) and validation (15%) sets. We, therefore, evaluated the performance of the proposed method using a test dataset that consists of the complete data from 2020 and is separated from the data used for training and validation.

5.3   Implementation Details

In this study, we used an Intel(R) Xeon(R) CPU @ 2.00 GHz and an NVIDIA Tesla T4 graphics processing unit (GPU) with 16 GB GDDR6 memory, 2560 shading units, 160 texture mapping units, and 64 render output units. The deep learning model was constructed using PyTorch 1.8 and the experimental results were analyzed using Matplotlib 3.1.3 and Seaborn 0.11.2. The mean computational time consumption and memory usages of the model training are 6.7 s and 1287.65 MB, respectively.

6   Results

The proposed DynamicDeepFlow approach involves a shallower learning model, k-means++, and a learning model, variational autoencoder (VAE). In this section, we analyzed the features learned by VAE and summarized the network traffic pattern clustering results by the k-means++.

6.1   Anomalous Network Traffic Pattern Identification

The objective of this study is to identify the anomalous network traffic pattern using the DDF approach. Figure 6 shows the model performance of the proposed DDF approach on network traffic pattern clustering for flow data in 2018, 2019, and 2020. The colors indicated the different clusters determined by the k-means++. Note the large circles referred to the cluster centers of different clusters. The k-means++ algorithm was used to classify the network traffic pattern to different clusters based on the flow change data. The rationale behind the k-means++ was the assumption that anomalous network traffic may form clusters that are different from the clusters of known network traffic patterns. We fed the high-dimensional features learned by the VAE to the k-means++. The clustering results of the k-means++ were decomposed into a 2-D dimension for visualization. Two significant observations can be made,

- First, years 2018 and 2019 and year 2020 had different number clusters. The network traffic pattern was categorized into five clusters for years 2018 and 2019 and four cluster for 2020, respectively. These results implied that a

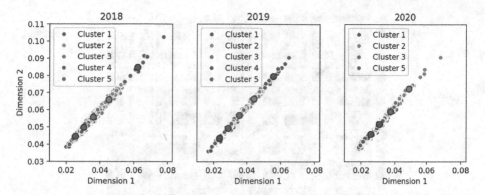

Fig. 6. Network traffic pattern clustering results by the proposed DDF approach for flow data in 2018, 2019, and 2020. The network traffic pattern was colored according to the clusters. (Color figure online)

different network traffic pattern in 2020 was identified. As shown in Fig. 6, the network traffic patterns located at the similar locations for the five different clusters of 2018 and 2019. Specifically, clusters had similar values of dimension 1 and dimension 2 except the cluster 5 of 2018, which had one network traffic pattern that was far away from it's cluster center. This network traffic pattern might be an outlier in terms of the cluster 5 of year 2018. On the other hand, cluster 4 (red) is missed in year 2020 and the network traffic patterns with high dimensions 1 and 2 were classified into cluster 2 (orange). It is note that the cluster center position of cluster 2 in 2020 was higher than it's in 2018 and 2019. Therefore, while year 2020 has one less cluster than years of 2018 and 2019, it is reasonable to say the cluster 2 of year 2020 was different with years 2018 and 2019.

– Second, to further investigate the anomalous network traffic patterns of 2020, we selected the orange point in the upper-right corner of Fig. 6 and visualized its Pearson correlation coefficient matrix in Fig. 7. This orange point represented the traffic network pattern on Wednesday night, February 5th, 2020. During this night, a strong sign of flow data exchange was evident as shown in Fig. 7. This strong flow data exchange sign was labeled an anomaly, since most of the flow data exchanges are likely to occur during the day rather than at night. Besides the selected orange point, we also investigated the other three orange points that had slightly lower dimensions 1 and 2 than the selected orange point in Fig. 6. The three orange points corresponded to Friday night, September 18th, Friday night, January 31st, and Wednesday, July 8th. These daytime and nighttime network traffic patterns were labeled as anomalies since cluster 2 was significantly lower in 2018 and 2019 than it was in 2020. By using a purely data-driven approach, we can determine the anomalies based on the clustering results. However, we are still uncertain about the implicit meaning of these anomalies. A solid explanation may be

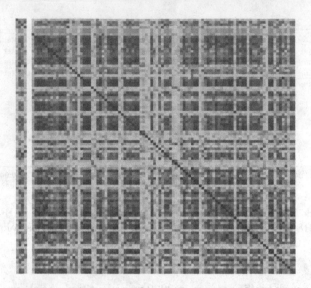

Fig. 7. An anomalous network traffic pattern.

found in understanding the reasons behind the anomalies based on physical knowledge of the network domain.

6.2   Sensitivity to Number of Cluster

It is commonly known that the model performance of k-means++ is closely related to the number of clusters. To identify the optimal number of clusters, we conducted a parametric study to empirically investigate the effect of the number of clusters on the SSE of k-means++. As can be observed from the zoom-in plot of Fig. 8, the sum of squared error (SSE) decreased as the number of clusters increased. The zoomed plot of Fig. 8 shows that the SSE with the increased number of clusters nearly converged to an asymptote at the cluster of 5. Therefore, the optimal number of clusters attempted for the k-means++ was set to 5 and this clustering category seemed to ensure satisfactory performance of the proposed DDF approach. The results described earlier (Fig. 6 and Fig. 7) were derived using this chosen value of 5. To quantify the variation of cluster center initialization, ten different initialization setting of cluster centers was conducted, and the results were used to plot the error bars showing the sensitivity of the proposed approach to the clusters.

6.3   Visualization of VAE Features

The VAE identified the anomalous network traffic pattern based on the latent vectors that were sampled from multivariate Gaussian distribution with parameters of μ and σ. The quality of these parameters significantly affected the model

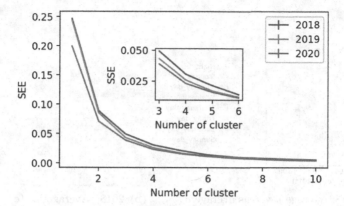

Fig. 8. SSE vs. number of clusters. Ten different initialization of cluster centers were conducted to plot the error bars showing the sensitivity of the k-means++ to the number of clusters.

performance in clustering. To understand the implicit meaning of these two parameters, we embedded the μ and σ in a two-dimensional space by averaging the feature of the μ and σ, respectively [20]. To uncover the possible differences in day and night and weekday and weekend, we classified the μ and σ into the night, Monday_day, Tuesday_day, Wednesday_day, Thursday_day, Friday_day, Saturday_day, and Sunday_day. Figure 9 shows the embedding results of the day, night, weekday, and weekend for the network traffic pattern of 2018, 2019, and 2020, respectively. Each figure created a scatter plot of average σ against average μ and two plots at the top and right margins. The top and right plots show the marginal distribution of the average μ and average σ, respectively. It is helpful to show these three plots together since the top and right plots summarized the pattern for every single variable and the scatter plot explored the relationship between these two variables and also described the strength of their relationship.

Based on these results, three important observations can be made and are listed as follows,

– There was a positive relationship between the average μ and average σ for daytime regardless of the years. It can be observed from the scatter plots (Fig. 9(a), Fig. 9(d), and Fig. 9(g)) that the average σ increased as the average μ increased. The strength of this relationship appeared to be strengthened as the year increased since the daytime data points got more concentration. On the other hand, the nighttime data displayed a moderate linear relationship between average μ and average σ, especially for 2019. Some nighttime points were far away from the rest of the nighttime points. While some nighttime points were far away from the rest of the nighttime points in 2018 and 2020, the number of them was negligible and could be recognized as the outliers. Overall, the three scatter plots displayed a clear linear relationship between the average μ and average σ for daytime regardless of the years.

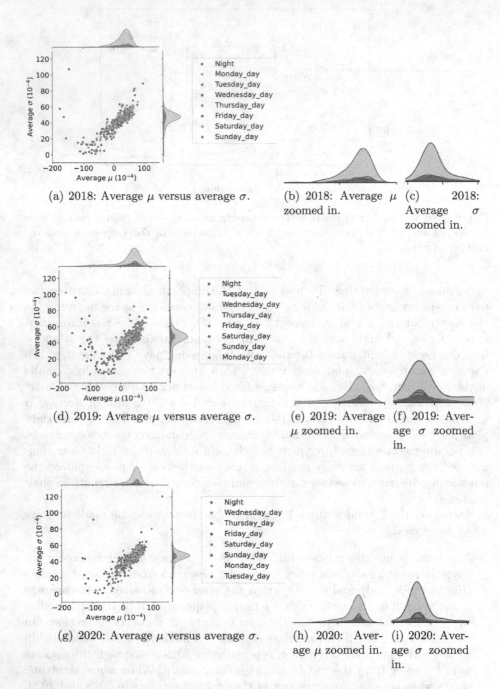

(a) 2018: Average μ versus average σ.

(b) 2018: Average μ zoomed in.

(c) 2018: Average σ zoomed in.

(d) 2019: Average μ versus average σ.

(e) 2019: Average μ zoomed in.

(f) 2019: Average σ zoomed in.

(g) 2020: Average μ versus average σ.

(h) 2020: Average μ zoomed in.

(i) 2020: Average σ zoomed in.

Fig. 9. Visualizations of the learned features by VAE for 2018, 2019, 2020. (Color figure online)

- The average μ and average σ got concentration as the year increased. The top and right plots showed that the distributions of average μ and average σ were wider in 2018 and 2019 than in 2020. While 2019 had a wider nighttime distribution compared to its in 2018, the left-skewed of 2019 might result from some outliers. The Friday_day (red) and Saturday_day (purple) got more concentration as the year increased. The concentration indicated (1) the network traffic pattern on Friday day and Saturday day had fewer changes, and (2) more similar the network traffic pattern happened on Friday and Saturday of 2020 than of 2019 and 2018.
- The network traffic pattern of daytime varied faster in 2018 and 2019 than in 2020. It can be observed from the scatter plots that the distribution of daytime was more dispersed in 2018 and 2019 than in 2020. This dispersed distribution of daytime indicated the more dynamic network traffic pattern changes in 2018 and 2019.

7   Conclusion

In this study, we propose a network traffic clustering approach, named DynamicDeepFlow (DDF), based on the flow changes in a dynamic network environment. The DDF is capable of approximating the complex data-to-traffic pattern relationship of a network from rapid change flow data. To the best of our knowledge, this is one of the first attempts to apply a real-time network traffic clustering approach to monitor network operations. We evaluated the effectiveness of the DDF using real network traffic data from the Energy Sciences Network. The evaluation results show that the proposed method can explicitly learn the dynamic nature of spatiotemporal traffic patterns, triggering anomalies in the coming days.

Under the assumption that network topology does not change, the proposed network traffic clustering method has shown plausible benefits for network operation. However, the network topology often changes to meet the growing demand for network resources. This limitation motivates us to explore further improvements to the proposed method. Our future study will investigate an adaptive clustering model to cope with changing network topologies.

References

1. Akyildiz, I.F., Lee, A., Wang, P., Luo, M., Chou, W.: Research challenges for traffic engineering in software defined networks. IEEE Netw. **30**(3), 52–58 (2016)
2. Fukuda, K., Takayasu, H., Takayasu, M.: Spatial and temporal behavior of congestion in internet traffic. Fractals **7**(01), 23–31 (1999)
3. Mallick, T., Kiran, M., Mohammed, B., Balaprakash, P.: Dynamic graph neural network for traffic forecasting in wide area networks. arXiv preprint arXiv:2008.12767 (2020)
4. Joshi, M., Hadi, T.H.: A review of network traffic analysis and prediction techniques. arXiv preprint arXiv:1507.05722 (2015)

5. Welzl, M.: Network Congestion Control: Managing Internet Traffic. Wiley, Hoboken (2005)
6. Kettimuthu, R., Vardoyan, G., Agrawal, G., Sadayappan, P.: Modeling and optimizing large-scale wide-area data transfers. In: 14th IEEE/ACM International Symposium on Cluster, Cloud and Grid Computing, pp. 196–205. IEEE (2014)
7. Uceda, V., Rodríguez, M., Ramos, J., García-Dorado, J.L., Aracil, J.: Selective capping of packet payloads for network analysis and management. In: Steiner, M., Barlet-Ros, P., Bonaventure, O. (eds.) TMA 2015. LNCS, vol. 9053, pp. 3–16. Springer, Cham (2015). https://doi.org/10.1007/978-3-319-17172-2_1
8. Singh, H.: Performance analysis of unsupervised machine learning techniques for network traffic classification. In: 2015 Fifth International Conference on Advanced Computing & Communication Technologies, pp. 401–404. IEEE (2015)
9. Soule, A., Salamatia, K., Taft, N., Emilion, R., Papagiannaki, K.: Flow classification by histograms: or how to go on safari in the internet. In: Proceedings of the Joint International Conference on Measurement and Modeling of Computer Systems, pp. 49–60 (2004)
10. Auld, T., Moore, A.W., Gull, S.F.: Bayesian neural networks for internet traffic classification. IEEE Trans. Neural Netw. **18**(1), 223–239 (2007)
11. Roughan, M., Sen, S., Spatscheck, O., Duffield, N.: Class-of-service mapping for QoS: a statistical signature-based approach to IP traffic classification. In: Proceedings of the 4th ACM SIGCOMM conference on Internet measurement, pp. 135–148 (2004)
12. Ahmed, M., Mahmood, A.N.: Novel approach for network traffic pattern analysis using clustering-based collective anomaly detection. Ann. Data Sci. **2**(1), 111–130 (2015)
13. Benesty, J., Chen, J., Huang, Y., Cohen, I.: Pearson correlation coefficient. In: Cohen, I., Huang, Y., Chen, J., Benesty J. (eds.) Noise Reduction in Speech Processing. STSP, vol. 2, pp. 1–4. Springer, Heidelberg (2009). https://doi.org/10.1007/978-3-642-00296-0_5
14. Shen, S., Sadoughi, M., Li, M., Wang, Z., Hu, C.: Deep convolutional neural networks with ensemble learning and transfer learning for capacity estimation of lithium-ion batteries. Appl. Energy **260**, 114296 (2020)
15. Shen, S., Sadoughi, M., Chen, X., Hong, M., Hu, C.: A deep learning method for online capacity estimation of lithium-ion batteries. J. Energy Storage **25**, 100817 (2019)
16. Li, T., Pasternack, G.B.: Revealing the diversity of hydropeaking patterns by time-series data mining. AGU Fall Meet. Abstr. **2020**, H049–03 (2020)
17. Li, T., Pasternack, G.B.: Revealing the diversity of hydropeaking flow regimes. J. Hydrol. **598**, 126392 (2021)
18. Shen, S., Sadoughi, M., Hu, C.: Online estimation of lithium-ion battery capacity using transfer learning. In: IEEE Transportation Electrification Conference and Expo (ITEC), pp. 1–4. IEEE (2019)
19. Shen, S., Sadoughi, M., Chen, X., Hong, M., Hu, C.: Online estimation of lithium-ion battery capacity using deep convolutional neural networks. In: International Design Engineering Technical Conferences and Computers and Information in Engineering Conference, vol. 51753, p. V02AT03A058. American Society of Mechanical Engineers (2018)
20. Shen, S., et al.: A physics-informed deep learning approach for bearing fault detection. Eng. Appl. Artif. Intell. **103**, 104295 (2021)

Unsupervised Anomaly Detection Using a New Knowledge Graph Model for Network Activity and Events

Paulo Gustavo Quinan[1], Issa Traore[1(✉)], Ujwal Reddy Gondhi[1], and Isaac Woungang[2]

[1] ECE Department, University of Victoria, Victoria, BC, Canada
{quinan,ujwalreddyg}@uvic.ca, itraore@ece.uvic.ca
[2] Department of Computer Science, Ryerson University, Toronto, ON, Canada
iwoungan@ryerson.ca

Abstract. The activity and event network (AEN) is a new knowledge graph used to develop and maintain a model for a whole network under monitoring and the relationships between the different network entities as they change through time. In this paper, we show how the AEN graph model can be used for threat identification by introducing an unsupervised anomaly detection model that leverages the probabilistic characteristics of the graph and the bits of meta rarity metric. A series of statistical features and underlying distributions are computed based on the graphical model of network activity and events. The anomaly scores of events are calculated by applying the bits of meta rarity to the aforementioned feature model and underlying distributions. Experimental evaluation is conducted a public cloud-based IDS yielding encouraging performance results.

Keywords: Anomaly detection · Graph database · Unsupervised machine learning · Intrusion detection system

1 Introduction

Intrusion detection systems (IDS) are broadly categorized into signature-based methods also called misuse detection and anomaly detection methods [6]. Signature-based detection consists of detecting attacks using characteristics of previously seen malicious events. Anomaly detection relies on the assumption that events that stray from normal system or user behavior are potentially malicious. While signature-based detection is good at detecting known attack patterns, by design they are not effective when confronted with new and unseen attack instances. In contrast, anomaly detection has the potential to detect novel attack instances. However, anomaly detection may involve a large number of false positives due to the fact that many atypical events are not necessarily malicious [2, 6].

We propose in this paper an unsupervised anomaly detection scheme based on a new knowledge graph called Activity and Event Network Model (AEN). The AEN framework uses large dynamic uncertain graphs to model and observe an interrelated network of

É. Renault et al. (Eds.): MLN 2021, LNCS 13175, pp. 117–130, 2022.
https://doi.org/10.1007/978-3-030-98978-1_8

activities and events over a period of time and across a broad set of hosts, and identify known and hidden attack patterns [8].

A series of statistical features and underlying distributions are computed based on a graphical model of network activity and events. The anomaly scores of events are calculated by applying the bits of meta rarity metric introduced by Ferragut et al. [3] to the aforementioned feature model and underlying distributions. The experimental evaluation based on the ISOT Cloud IDS (ISOT-CID) dataset [1] yields encouraging results in terms of detection accuracy.

The rest of the paper is structured as follows. Section 2 discusses related work. Section 3 gives a brief overview of the AEN graph model. Section 4 presents the proposed AEN-based anomaly detection model. Section 5 presents the experimental evaluation of the proposed detection model. Finally, Sect. 6 makes concluding remarks.

2   Related Work

Traditional rule-based IDSs are very capable of identifying and blocking obvious threats and known attacks but have very little chance of identifying novel attack patterns, which sometimes are multi-stage, custom crafted, attack vectors. That resulted in the development of anomaly-based IDSs, which are capable to detect novel attack types but suffer from high level of false positives and are still incapable of effectively connecting the dots among intrusions. Sommer and Paxton [6] argue that a mismatch between the capabilities of machine-learning and the type of analysis required for intrusion detection has limited the improvement of those systems and that limitations in the descriptive capabilities of these tools have curtailed the understanding of alerts by security analysts creating a "semantic gap" which further limits their adoption. The authors also propose a set of guidelines for applying machine learning to network intrusion detection.

Even so, more and more research is being performed in anomaly, misuse or behavioural based IDSs given the limitation of rule-based IDSs in identifying novel threats. This trend is confirmed by Luh et al. [5]'s extensive review of works in advanced persistent threat (APT) detection. The review provides several other interesting findings. For instance, it shows an almost even split between malware detection, host intrusions and network intrusions. That is explained by the multi-modal nature of APT attacks, which require a varied set of equally important tools, each covering different intrusion stages. The paper also shows that the majority of the works are focused on a single area of detection, be it host-based or network-based, with only a few utilizing multiple sources of data. That demonstrates a gap in the literature given what was discussed in the previous paragraph and the importance of linking data from multiple sources to identify hidden patterns of compromise among them.

Nevertheless, many novel APT detection and prevention techniques have been proposed or developed in the past few years.

Giura and Wang [4] proposed a new anomaly-based detection framework, conceptualized on the attack pyramid model created by the same authors. The system uses data from several different network and application logs to categorize intrusion events into the pyramid's planes (life cycle phases) and correlate events across the planes to detect sophisticated attacks such as APTs.

3   AEN Graph Model Overview

The AEN graph model is a new graph framework developed at the ISOT Lab to capture activities and events occurring in a network over time [8]. The AEN graph captures the uncertainty and dynamicity of network activity using a random time-varying multigraph model. Figure 1 shows a partial example of the AEN graph model.

The data available in the ISOT-CID dataset [1] contains files in multiple formats, ranging from memory logs, TCP packet captures, syscall logs, from which the AEN graph is generated.

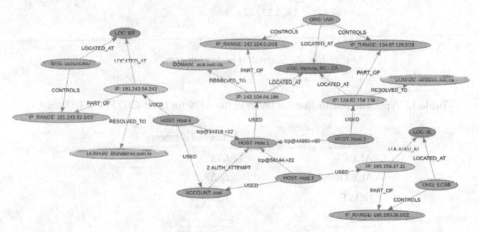

Fig. 1. Sample AEN graph

The AEN model provides the foundation for detecting, responding and forensically investigating current and new attack vectors using data from both the traditional security ecosystem and beyond the organization perimeter. The AEN graph is a multigraph with a variety of nodes and edges.

Figure 2 and Table 1 give an overview of AEN nodes. As shown in Fig. 2 every node element has an id, a label and corresponding properties. The label represents the type of node.

```
[{'id': 3505522295267017419,
  'label': 'ALERT',
  'properties': {'destIP': '172.16.1.24',
  'protocol': 'tcp',
  'sourcePort': 22,
  'destPort': 37136,
  'sourceIP': '142.104.64.196',
  'service': '',
  'classification': '',
  'priority': 0,
  'timestamp': '2016-12-09T17:49:59.205766Z'}},
 {'id': -79037162220219263361,
  'label': 'ALERT',
  'properties': {'destIP': '172.16.1.24',
  'protocol': 'tcp',
  'sourcePort': 16783,
  'destPort': 23,
  'sourceIP': '14.162.253.53',
  'service': '',
  'classification': '',
```

Fig. 2. Snippet of node elements.

Table 1. Types and distributions of nodes in the AEN for ISOT CID Phase 1 Dataset

Node label	Number of elements
ALERT	12,672
ACCOUNT	606
HOST	252
DOMAIN	252
IP	252
IP_RANGE	195
ORGANIZATION	91
LOCATION	81
Total	**14,401**

Figure 2 shows the data for two AEN nodes, with "ALERT" label assigned to them, and the corresponding properties. Table 1 shows the types of AEN node elements present in ISOT CID Phase I, along with their distributions.

Figure 3 and Table 2 give an overview of AEN edges. Each edge element has an id, a label and corresponding properties. Figure 3 shows data for two edges with "SESSION" label or type assigned to them and the corresponding properties.

```
[{'id': 5712465223099226337,
    'label': 'SESSION',
    'properties': {'__maliciousLabel': False,
    'destSize': 52,
    'protocol': 'tcp',
    'sourcePort': 38040,
    'destPort': 22,
    'packetCount': 11,
    'fragmentedPacketCount': 0,
    'deltaTime': 0,
    'tcpState': 6,
    'startTime': '2016-12-16T17:18:31.940Z',
    'stopTime': '2016-12-16T17:18:31.962Z',
    'sourceSize': 22768},
    'source': -4539149150930014966,
    'destination': -5938715740280398657},
{'id': 2267145363307886547,
    'label': 'SESSION',
    'properties': {'__maliciousLabel': False,
    'destSize': 0.
```

Fig. 3. Snippet of edge elements

Table 2. Types and distributions of edges in the AEN for ISOT CID Phase 1

Edge label	Number of elements
AUTH_ATTEMPT	262,258
SESSION	115,451
ALERT_TRG_BY_HOST	12,672
HOST_USED_IP	252
IP_RES_DOM	252
IP_LOCATED_AT	252
IP_PART_RANGE	252
ORG_CTRL_RANGE	196
ORG_LOC_AT	119
Total	**391,704**

Table 2 shows the different types of edge elements present in the ISOT CID Phase I data along with their distributions.

4 Proposed Anomaly Detection Model

4.1 Measure of Anomalousness

Most available unsupervised anomaly detection approaches use statistical methods to cluster the data closer to each other and look out for outliers/rare events to flag them as anomalous events. Anomaly is defined as something that deviates from what is standard, normal, or expected [2].

Tandon and Chan [7] defined bits of rarity given by formula (1) as a measure to calculate the anomaly score of an event. Given a probability distribution described by probability density or probability mass function f, the anomaly score of an event x is given by Eq. (1).

$$R_f(x) = -\log_2 P_f(x) \tag{1}$$

Ferragut et al. [3] suggests that the definition of what is an anomaly is subjective to the type of data that is being worked on, and that definition is not comparable among two different types of data if presented, so they proposed the definition of anomalousness based on the probability of the probability of the event rather than just the probability of an event. They define the Bits of Meta-Rarity as a measure to calculate the anomaly score of an event that is *regulatable* and *comparable*, defined as follows. Given a (discrete or continuous) random variable X with probability density or mass function f defined on the domain D, let $A_f : D \to \mathbb{R}_{\geq 0}$

$$A_f = -\log_2 P_f(f(X) \leq f(x)) \tag{2}$$

Example 1: Given, three events, A, B and C with individual properties and respective probabilities p(A), p(B) and p(C), such that $p(A) < p(B) < p(C)$, then the anomaly score of these events is given by

$$Anomaly\ score(A) = -\log_2(p(A))$$

$$Anomaly\ score(B) = -\log_2(p(A) + p(B))$$

$$Anomaly\ score(C) = -\log_2(p(A) + p(B) + p(C))$$

We assume negative log of zero to be positive infinity by convention. We can see that as the probability of the event increases the anomaly score of the event decreases. In other words, the event becomes more common, i.e. less rare or less anomalous. By calculating the bits of meta rarity we are calculating the rarity of the rarest events, in the given sample.

In our work, since we are dealing with large graphs, the sample size is huge, and the feature space is very spatial, and as a result the following issue arose.

Example 2: Imagine a case wherein we are calculating *feature1* as the number of unique sessions between Source S and destination D, and assume we get the following values out of *feature1*:

$$A_1 - B_1 - 22$$

$$A_2 - B_1 - 11$$
$$A_3 - B_1 - 22$$
$$A_4 - B_2 - 555$$
$$A_5 - B_6 - 10$$
$$N_3 - A_7 - 9$$

On a simple view of the values of *feature1* extracted for the given source/destination pairs, we assume the anomaly score of A_4-B_2 should be higher. But as we calculate using the current anomaly model, we first consider which of these tuples have a high frequency of occurrence, in which case we can see that A_1-B_1 and A_3-B_1 have the same number of sessions, hence these two sessions have a higher probability (i.e. 2/6) than the others resulting in low anomaly score for (A_1-B_1, A_3-B_1). Whereas tuples A_2-B_1, A_4-B_2, A_5-B_6 and N_3-A_7 appear only once and have the same low probability (i.e. 1/6) resulting in the same anomaly score for these 4 tuples. If we consider just the abovementioned 4 tuples, then A_4-B_2 will look as anomalous as the 3 others because of having the same probability. However, intuitively this should not be the case as the number of sessions in A_4-B_2 is farther from the number of sessions in the other tuples.

Hence to overcome this issue we segregate the tuples into bins, so that all the above tuples other than 555 belong to one bin and they can have high probability (i.e. 5/6) resulting in a low anomaly score.

To achieve the concept of bin implementation, we transform the features space by rounding it according to the upper bound of the bin where the value lies. For example, if the bin size is 5 and the feature is 22, the value will be rounded to 25.

In our feature model, which is described later, each feature is defined as a tuple (e.g. 1-tuple, 2-tuple, 3-tuple, etc.). Over a time frame ΔT, we compute for each feature type a list of tuples $u_i = \{(a_{i1}, v_{i1}), (a_{i2}, v_{i2}), \ldots\}$, where each tuple consists of a feature a_{ij} and its value v_{ij}. For instance, in our feature model, the feature *uniquesessions* is represented by the tuple $a_{ij} = (Source\,IP, Destination\,IP)$ and the number of unique sessions between these IPs as the value v_{ij}.

We discretize the feature values by binning. For each tuple, this involves rounding the feature value to the upper bound of the corresponding bin.

For a given feature type a, assume that we obtain the following sequence of bins $b_1...,b_n$ over the time frame ΔT where n is the number of bins. The empty bins are ignored. The probability $p(b_i)$ for bin b_i is defined as the ratio of the number of tuples in the bin divided by the total number of tuples over all non-empty bins as follows:

$$p(b_i) = \frac{|b_i|}{\sum_{j=1}^{n} |b_j|}$$

where $|x|$ denote the size of bin x.

Given a bin b_m, the anomaly score of the tuples belonging to b_m is given by

$$A_{b_m} = -log_2 \left(\sum_{(\forall i : p(b_i) \leq p(b_m))} p(b_i) \right)$$

Later, the anomaly score of the tuples in the feature list are compared to the pre-set threshold, and if the anomaly score of an element is greater than the threshold, we generate an alert.

4.2 Features Model

Our feature model involves a wide range of features extracted from the AEN graph elements. We describe in the following these features under the categories of features derived from session data and features based on authentication data.

4.2.1 Feature Extraction from Sessions Data

One of the main type of edges in the AEN graph model is the session edge. Our proposed detection model leverages session edges to identify useful features that can be used for effective threat identification.

A snippet of a sample session edge data is depicted in Fig. 4. The snippet contains the attributes of the session edge and corresponding values.

```
{'id': 2808529043190439159,
 'label': 'SESSION',
 'properties': {'destSize': 4350,
 'protocol': 'tcp',
 'sourcePort': 22,
 'destPort': 34748,
 'packetCount': 13,
 'fragmentedPacketCount': 0,
 'deltaTime': 0,
 'tcpState': 2,
 'startTime': '2016-12-15T18:46:53.156Z',
 'stopTime': '2016-12-15T18:46:53.198Z',
 'sourceSize': 543},
 'source': -5938715740280398657,
 'destination': -4539149150930014966}
```

Fig. 4. A snippet of session edge data

Post correlating with the source and destination hosts, we get the source and destination IP addresses which are associated with the source and destination nodes corresponding to the edge.

We extract 9 different feature types from session data. These feature types are described in the following.

- *Unique sessions:* consists of the total number of sessions between $host_1$ and $host_2$, in the given time period for each pair of hosts $host_1$ and $host_2$ in the data.
- *Unique destination ports:* consists of the number of unique ports of host h_2 that are being accessed by host h_1 in a given time frame ΔT.

- *Unique destinations with same destination port*: corresponds to the number of unique hosts h_2 to which host h_1 connects with the same destination port in a given time frame.
- *Number of destination ports for a source host*: is obtained by calculating the number of unique destination ports the host h1 is trying to access in a given time frame.
- *Time between sessions*: is computed as the mean time duration between the termination of a session and the start of the next session for a given source and destination.
- *Length of sessions*: is computed as the average length of the session duration from a given source to destination over a time frame ΔT. This gives us the amount of data that could have been trickled out from the server. Varying the length of the time frame over long periods could help detect advanced persistent threats (APTs).
- *Ratio of session size*: corresponds to the average ratio of the destination size to the source size for a given source and destination tuple over a time frame.
- *Session velocity:* is obtained as the ratio of the number of packets in a session to the duration of the session. This gives us an idea of the velocity of the packet transfer between source and destination expressed in packets/sec.
- *Session source size:* is computed as the average of the source size of the packets sent from a unique source IP to all the destinations during a given timeframe ΔT.

4.2.2   Feature Extraction from Authentication Data

The authentication data modeled by the AEN graph is a potential source of useful information toward anomaly detection. We extract 6 different features from the authentication data captured by the AEN model.

- *Unique authentication failures:* is computed as the total number of failed authentication attempts from host h1 to host h2 in the given time frame ΔT. This includes the number of failed attempts regardless of whether it was the password or username that was incorrect, or even failed public key logins as well.
- *Unique usernames:* is computed as the number of unique usernames that were used in failed authentication attempts by the host h1 to destination h2 within a given time frame ΔT.
- *Unique source username:* computes the number of times a unique failed username is used by host h1 to all other destinations in a given time frame ΔT.
- *Unique source destination username:* is computed as the number of unique source IP, destination IP and failed usernames in a given time frame ΔT.
- *Username dispersion per source:* computes the number of failed unique usernames used by the source system h1 to all other destinations in a given time frame ΔT.
- *Username stuffing:* is computed as the number of unique destinations, a source h1 is trying to access with the same username.

5   Experimental Evaluation

In this section, we give an overview of the ISOT cloud IDS evaluation dataset and then provide and discuss the evaluation results of the proposed anomaly detector.

5.1 Dataset

The dataset used in our work was collected previously by the Information Security and Object technology (ISOT) lab of the University of Victoria for cloud intrusion detection system design and evaluation. The ISOT Cloud intrusion detection dataset (ISOT CID) is available publicly. It is a labeled dataset which consists of over 8 terabytes of data, involving normal activities and a wide variety of attack vectors, collected in two phases (phase 1 in 2016 and phase 2 in 2018) and over several months for the VM instances and several days and time slots for the Hypervisors.

The benign/normal data are from web applications and administrative activities ranging from maintaining the status of VMs, rebooting, updating, creating files, SSH'ing to the machine, and logging in a remote server.

The web traffic was generated by more than 160 legitimate visitors, including more than 60 human users and genuine traffic generated by 100 robots, performing tasks such as account registration, reading/posting and commenting on blogs, browsing various pages, and so on. One of the web applications consist of a password management service for registered users.

The cloud environment had 3 hypervisor nodes, A, B and C, but only A and B were part of the attack.

Table 3. ISOT CID data structure [1, 2]

VM Label	Operating System	Zone / Node	Internal / External IP Address
VM 1	Centos	C	172.16.1.10 / 206.12.59.162
VM 2	Centos	A	172.16.1.28
VM 3	Debian	A	172.16.1.23 / 206.12.96.142
VM 4	Windows Server 12	A	172.16.1.26 / 206.12.96.149
VM 5	Ubuntu	A	192.168.0.10
VM 6	Centos	A	172.16.1.19 / 206.12.96.240
VM 7	Ubuntu	B	172.16.1.20 / 206.12.96.239
VM 8	Ubuntu	B	172.16.1.24 / 206.12.96.143
VM 9	Windows Server 12	B	172.16.1.21 / 206.12.96.141
VM 10	Centos	B	172.16.1.27 / 206.12.96.150

Table 3 depicts the distribution of virtual machines among the A, B, and C Zones. These systems are assumed to be benign unless compromised and used to attack other machines. The ISOT CID involves a wide variety of logs, e.g. network traffic logs, system logs, events logs, CPU/Disk utilization, system calls, etc.

Figure 5 shows the distribution of the network traffic in the phase I of ISOT-CID.

Total Normal Traffic	Total malicious Traffic	Total Packets
22356769	15649	22372418
99.93%	0.07%	100%

Date	Node A			Node B		
	Total Normal Traffic	Total Malicious Traffic	Total Packets	Total Normal Traffic	Total Malicious Traffic	Total Packets
Day 1 9/12/16	3294811	982	3295793	2975278	243	2975521
Day 2 15/12/16	6704732	6833	6711565	4832628	1174	4833802
Day 3 16/12/16	522590	1470	524060	1549115	618	1549733
Day 4 19/12/16	1419617	3667	1423284	1057998	662	1058660

Fig. 5. ISOT CID Phase 1 data distribution

5.2 Performance Evaluation

The performance evaluation of the proposed anomaly detection model was conducted using phase I of the ISOT CID dataset.

Table 4. Phase 1 IP addresses distribution

Total unique source IPs	335
Total malicious IPs	189
Total benign IPs	146

Table 4 shows the distribution of unique IP addresses involved in the ISOT CID phase I dataset.

The experiment involved constructing the AEN graph based on the dataset and extracting the features from the corresponding AEN graph. Then, the anomaly scores were computed and compared against a threshold, resulting in flagging the samples as either intrusive or legitimate. By comparing the flags against the original labels (in the dataset), we identified the false positives and false negatives.

The following parameters were varied as shown below:

- Bin size – [1,2,3,4,5,10,20]

- Time Frame ΔT – [15 min, 60 min, 240 min, 1440 min]
- Threshold – varied from 2 through 21

For the chosen parameters, we calculate the detection rates and false-positive rates which are defined as follows.

Detection Rate: The detection rate is defined as the number of malicious IPs detected by the model to the total number of malicious IPs in the data.

$$Detection\ rate = \frac{Malicious\ IP's\ detected}{Total\ number\ of\ malicious\ IP's} \tag{5.1}$$

False-Positive Rate: The false-positive rate is defined as ratio of the number of benign IPs that have been reported as malicious to the total number of benign IPs

$$False\ positive\ rate = \frac{Number\ of\ regular\ IP's\ mislabelled\ as\ malicious}{Total\ number\ of\ regular\ IP's} \tag{5.2}$$

The performance (DR, FPR) tuples of the proposed detection model were calculated by varying the bin size, the time frame and the threshold parameter values based on ISOT CID phase 1.

By analyzing the results, the following parameter combination was found to be yielding the best detection rate/false positive rate combination:

– Timeframe = 15 min
– Bin size = 3
– Threshold = 3

The detection rate, with the above parameters is 85.19%, with a false positive rate of 17.12%, and the next best performance is a detection rate of 79.89% and False positive rate of 15.75%, with the following parameter combination.

– Timeframe = 15 min
– Bin size = 4
– Threshold = 3

By analyzing the false positives, it was found that most of the IPs that have been alerted as malicious are due to the nature of the data distribution. It was found that some benign IPs consisted of a large percentage of the benign traffic, and hence have been alerted by the anomaly detection model, which is desirable, but not in this data distribution. One solution that can be implemented to suit the current dataset is to use a hybrid model, which is an anomaly detection model, but has some ground truth associated as well, so that such false positives can be reduced.

6 Conclusion

Though the number of signature based detection models and secure infrastructure are increasing every day, so are the number of cyber security breaches both in volume and impact on the organizations. This warrants the need for more unsupervised anomaly detection models, which could detect zero day attacks and if cascaded with the signature based detection models, could really reduce the associated risk while reducing alert fatigue.

This paper presents an unsupervised anomaly detection scheme based on the AEN graph, which is a large graph model used to capture the activity and events underlying organization network. The features used, along with the method of extracting features and calculating the anomaly scores of events are discussed. The issue concerning the spatial feature space is discussed along with feature transformation method to resolve the spatial nature of the feature data.

The experimental evaluation of the detection model on the phase-1 data of the ISOT Cloud IDS dataset yields a performance consisting of a detection rate of 85.19% with a false positive rate of 17.12%.

Unsupervised detection is desirable but very challenging. On the first hand, it is beneficial because it doesn't require prior knowledge of the environment or threat tactics; no training set or training period is required. Hence, it has a much stronger potential to detect novel attack patterns compared with supervised approaches and it represents a better fit for real-world operational environments and constraints. On the other hand, it is challenging because of the lack of prior knowledge used in threat decision-making. Therefore, the performance achieved by our models should be assessed in this context. Our goal is to continue working on improving the detection performance of AEN-based threat detection by expanding the feature space and investigating and developing further threat detection schemes. The feature space presented in this paper is only a subset of what is possible. More features can be derived from the AEN graph model and help with improving detection accuracy. Also, by combining different schemes, we expect to maximize threat detection coverage and accuracy. We plan to report the IP's as malicious if two or more features report that the specific source IP as malicious instead of reporting all the Source IPs alerted by any of the features.

As indicated above, the proposed threat detection approach was evaluated using of ISOT-CID Phase I. We are working on extending the evaluation to using both ISOT-CID Phase II and a related dataset captured by the University of New Brunswick (UNB).

Future work will also deal with the implementation of the proposed unsupervised anomaly detection model cascaded with a signature-based model to read data from the AEN graph database automatically for every preset time frame and performs the anomaly detection.

References

1. Aldribi, A., Traore, I., Moa, B.: Data sources and datasets for cloud intrusion detection modeling and evaluation. In: Mishra, B.S.P., Das, H., Dehuri, S., Jagadev, A.K. (eds.) Cloud Computing for Optimization: Foundations, Applications, and Challenges. SBD, vol. 39, pp. 333–366. Springer, Cham (2018). https://doi.org/10.1007/978-3-319-73676-1_13

2. Aldribi, A., Traore, I., Moa, B.: Hypervisor-based cloud intrusion detection through online multivariate statistical change tracking. Comput. Secur. **88**, 101646 (2020)
3. Ferragut, E., Laska, J., Bridges, R.: A new, principled approach to anomaly detection. In: Proceedings of the 2012 11th International Conference on Machine Learning and Applications (ICMLA 2012) (2012)
4. Giura, P., Wang, W.: A context-based detection framework for advanced persistent threats. In: 2012 International Conference on Cyber Security, December 2012, pp. 69–74
5. Luh, R., Marschalek, S., Kaiser, M., Janicke, H., Schrittwieser, S.: Semantics-aware detection of targeted attacks: a survey. J. Comput. Virol. Hack. Techniq. **13**(1), 47–85 (2017)
6. Sommer, R., Paxson, V.: Outside the closed world: on using machine learning for network intrusion detection. In: Proceedings of the 2010 IEEE Symposium on Security and Privacy (SP 2010), Washington, 2010, pp. 305–316. IEEE Computer Society (2010)
7. Tandon, G., Chan, P.K.: Tracking user mobility to detect suspicious behavior. In: Proceedings of the 2009 SIAM International Conference on Data Mining, pp. 871–882 (2009)
8. Traore, I., Quinan, P.G., Yousef, W.: The activity and event network (AEN) model: graph elements and construction. Technical report, ISOT lab, ECE Department, University of Victoria (2020)

Deep Reinforcement Learning for Cost-Effective Controller Placement in Software-Defined Multihop Wireless Networking

Afsane Zahmatkesh⬤ and Chung-Horng Lung⬤

Department of Systems and Computer Engineering,
Carleton University, Ottawa, Canada
afsane.zahmatkesh@carleton.ca, chlung@sce.carleton.ca

Abstract. One of the key features in software-defined networking (SDN) with a multi-controller environment is the controller placement problem (CPP) that aims to find the number of controllers, controller placements and controller assignment. Solving the CPP in software-defined multihop wireless networking (SDMWN) has a significant impact on the generated control overhead. In SDMWN, devices use unreliable and shared multihop wireless communications without the help of any infrastructure, such as a base station or an access point. Various algorithms have been proposed to find near-optimal solutions for the CPP. Deep reinforcement learning (DRL) becomes popular in various fields and a few studies have also investigated using DRL for the CPP in wired networks and infrastructure-based wireless networks. However, DRL has not been researched for the CPP in SDMWN. Hence, in this paper, the potential of using DRL to the CPP in SDMWN for a given number of controllers is investigated to minimize the generated control overhead referred to as the network cost. The results show that the adapted DRL is able to find controller placements and assign the controllers to devices such that the obtained network cost and the average number of hops among devices and their assigned controllers, as well as the average number of hops among different controllers in the network, are close to those obtained from the optimal solutions.

Keywords: Software-defined multihop wireless networking (SDMWN) · Controller placement problem (CPP) · Control overhead · Deep reinforcement learning (DRL)

1 Introduction

In an SDN architecture with a multi-controller control plane, controller placement problem (CPP) [1–3] has gained a significant attention. The objective of the CPP is to find the number of controllers, controller placements and controller assignments while considering different objectives and constraints. Some

E. Renault et al. (Eds.): MLN 2021, LNCS 13175, pp. 131–147, 2022.
https://doi.org/10.1007/978-3-030-98978-1_9

of these objectives presented in the literature to solve the CPP in wired networks are the delay between network devices and their assigned controllers [1], load-balancing among controllers [4], control overhead [5] and the control plane deployment cost [6].

In software defined multihop wireless networking (SDMWN) [7], due to the unique characteristics of some types of networks, only one wireless interface or channel is often used to forward both data and control traffic. An example is wireless sensor networks (WSNs). Further, wireless communications may be unreliable and interference may occur for shared medium, especially for multihop communications. As a result, packet losses may happen more frequently and delay could be high, which could have a significant impact on network performance in terms of energy consumption, reliability, and quality. Solving the CPP in SDMWN plays an important role on the generated control overhead in the network. The control overhead is referred to as the network cost [5]. Selecting the right number of controllers, and processing controller placements and controller assignments influence the connectivity among controllers and devices, the network cost and delay in the control plane.

The CPP is a NP-hard problem [1]; therefore, formulating the CPP as an optimization problem to find the optimal solutions results in high computational complexity. Consequently, various methods and algorithms have been proposed to find near-optimal solutions to the problem [8]. Most models have some assumptions or have been linear to deal with the complexity. Modeling of SDMWN could become much more complicated if more details are considered. Hence, algorithm- or model-based approaches are not suitable for the CPP in SDMWN. On the other hand, machine learning (ML) [9] based approaches do not relay on models, but data. As a result, ML-based approaches are more suitable for complicated problems. Recently, deep reinforcement learning (DRL) [10,11] has gained increasing attention for different types of problems using the combination of deep learning [12] and reinforcement learning (RL) [13]. Deep learning, known as deep neural network, is a subset of machine learning inspired by the structure of the human's brain. In RL, using observations from the environment and a Q-table, an agent maps the situations in the environment to the actions to maximize the achieved reward. DRL is able to overcome the existing challenges of RL, including low-dimension of situation-action representations in Q-table based on the advantages of artificial neural networks. Using the training data, DRL is able to learn and make predictions for the new data sets, which is beneficial in dynamic environments [10,11].

Recently, a few research efforts have been conducted to apply RL and DRL to the CPP for wired networks [14,15] and the internet of vehicles (IoV) [16]. DRL also has the potential to solve the CPP in SDMWN considering the characteristics of SDMWN, including multihop wireless communications and dynamic topology changes. To our knowledge, both RL and DRL have not been be investigated for the CPP in SDMWN. Therefore, the objective of this paper is to investigate the feasibility and effectiveness of applying DRL to the CPP in SDMWN by considering specific characteristics of SDMWN.

In [5], both nonlinear and linear optimization problems are proposed to solve the CPP in SDMWN. Because of the high computational complexity of the nonlinear model, the objective of our paper is to adapt DRL to solve the simplified and linear optimization problem presented in [5] for the CPP in SDMWN while minimizing the network cost. The network cost is considered as the total number of control packets transmitted in the control plane to exchange the configurations, set up flow rules and discover the network topology. Specifically, to find the placements of N controllers and assign the controllers to the network devices, a deep Q-network (DQN) [17] is adapted for SDMWN in our proposed approach. DQN is able to overcome the aforementioned challenges in RL in an environment with a high-dimensional state-action space and to approximate the Q-value function for all possible actions using deep learning.

The first contribution of the paper is to model and solve the CPP for SDMWN with DQN. To our knowledge, solving the CPP for SDMWN has not been reported in the literature. Secondly, we compare the performance of our proposed DQN-based approach with the optimal solutions achieved from the linear optimization problem presented in [5] in terms of the network cost and the average number of controller-device hops as well as the average number of inter-controller hops.

The rest of the paper is organized as follows. Section 2 presents an overview of the related work in using RL and DRL in solving the CPP. Section 3 introduces the system model used in this paper. The evaluation of the proposed model is presented in Sect. 4. Finally, Sect. 5 concludes the paper and presents some future research directions.

2   Related Work

Recently, research works have been conducted to apply RL and DRL to the CPP. In [14], a heuristic algorithm based on learning automaton is proposed to find the number of controllers and controller placements in wired networks to minimize latency among controllers and devices. In this paper, using random actions and the evaluations of the actions, the learning automaton learns to select optimal actions in the network. The results show that the proposed approach is able to reduce latency compared to the K-means clustering algorithm [18].

In [15], the authors propose a multi-objective optimization problem (D4CPP) to solve the CPP in wired networks. Their aim is to minimize latency among the devices and their assigned controllers and balance the load of the control plane among controllers. D4CPP also considers data flow fluctuations in the network. Therefore, to solve the problem in a dynamic environment, a DQN algorithm based on the data flow fluctuations is proposed. D4CPP consists of two sub-systems: in the first sub-system, the objective is to find the placements of a given number of controllers; and the second sub-system then dynamically adjusts the controller assignments to network devices. The results demonstrate that, compared to the K-means algorithm [18], D4CPP is able to improve the network performance in terms of latency among the devices and controllers and load-balancing among controllers.

Yuan et al. in [16] propose a dynamic controller assignment in IoV with a hierarchical control plane using multi-agent DRL to minimize control delay and control traffic load among vehicles and their assigned controllers. In the proposed approach, multi-agent deep deterministic policy gradient (MADDPG) [19], the controller placements are fixed and act as agents in the algorithm to dynamically adjust controller assignments. The results show that the proposed approach is able to reduce packet loss in the network.

Based on the reviewed studies in the literature, applying DRL to the CPP can be beneficial in a wired network in terms of network performance metrics including minimizing delay and control traffic, and load-balancing in a network. In this paper, the objective is to investigate the potential and effectiveness of adapting DRL to solving the proposed CPP in SDMWN by considering SDMWN-specific characteristics with the aim of minimizing the network cost.

3   System Model

In this section, we present how to adapt DQN to the proposed linear CPP for SDMWN in [5] to minimize the network cost. In the proposed CPP model in [5], the objective is to find a given number of controller placements and assign the controllers to network devices that aims to minimize the network cost. The network cost includes the total number of controller-device control packets to exchange configurations, set up flow rules and discover the partial/local views of the network as well as the total number of inter-controller control packets to synchronize different controller's views and achieve a global view of the network. In the linear optimization model used in our paper, it is assumed that both controller-device and inter-controller communications use the shortest paths. In this case, the outputs of the linear optimization model consist of the controller placements, controller assignments to network devices and the minimum network cost.

To solve a problem using a DQN algorithm, we need to describe different parameters and definitions for SDMWN, which are listed as follows [10,11,17].

- **Environment:** The main part of a problem that needs to be defined in the DQN algorithm is the environment that generates the input data. The DQN algorithm is adapted in this paper to the CPP in SDMWN; hence, the environment is an SDMWN with a number of wireless network devices.
- **State Space (S):** It represents the set of possible situations in the environment. Equation (1) shows the set of states in the proposed problem in this paper. The set consists of the state of all network devices in a network with V devices such that s_i demonstrates the state of device i. As shown in Eq. (2), the state of device i consists of w_i^j for device j ($j \in V$) in the network. w_i^j demonstrates if device j is assigned to a controller placed on device i as shown in Eq. (3). $x_{i,j}$ equals one if and only if the device i is assigned to the controller placed on device j and each device is only assigned to one controller in the network.

$$S = \{(s_1, s_2, ..., s_{|V|})\} \tag{1}$$

$$s_i = \{w_i^1, w_i^2, ..., w_i^{|V|}\} \tag{2}$$

$$w_i^j = \begin{cases} 1, & if \ x_{i,j} = 1 \\ 0, & otherwise, \end{cases}, \forall i,j \in V,$$

$$s.t. \ \sum_{i=1}^{|V|} x_{i,j} = 1 \tag{3}$$

- *Action Space (A):* It represents the possible actions that can be taken in a defined environment. Using the current observation and DQN, a sequence of actions will be applied that can change the state of the environment. Equation (4) shows the action space in an environment that consists of the actions applied at various particular times. As shown in Eq. (5), the action at time t (a_t) consists of the state of all devices form 1 to $|V|$ at time t ($b_t^j, j \in V$). As demonstrated in Eq. (6), in the proposed algorithm in this paper, a possible action at a particular time demonstrates the network devices that will be selected as controllers. y_i equals one if and only if there is a controller placed on device i and the total number of controllers in the network is N. In this case, the action changes the current state of the environment and the selected controllers will be assigned to the devices in the network.

$$A = \{a_l, a_{l+1}, a_{l+2}...\} \tag{4}$$

$$a_t = \{b_t^1, b_t^2, ..., b_t^{|V|}\} \tag{5}$$

$$b_t^j = \begin{cases} 1, & if \ y_j = 1 \\ 0, & otherwise, \end{cases}, \forall j \in V,$$

$$s.t. \ \sum_{j=1}^{|V|} y_j = N \tag{6}$$

- *Agent:* Another essential part of DQN is an agent that is responsible for making decisions and taking a sequence of actions in the environment. Based on [15], in the proposed DQN in this paper, it is assumed that the agents are devices.
- *Reward (R):* When the agent takes an action, the environment obtains a reward. R shows the set of rewards obtained using DQN during run time. In the proposed DQN, the reward is the network cost of the possible solution demonstrated by the action and the new states.
- *Transition:* In DQN, when an agent takes an action a at a particular time in the environment, a transition from state s to state s' happens in the environment that obtains reward r.
- *Q-function:* Using the current state in the environment, DQN approximates the action-value function referred to as a Q-function that shows the reward of all possible actions using the concept of neural networks.

- **Replay Memory:** In DQN, during run time, the transitions will be saved in a replay memory. After that, by sampling randomly stored transitions called mini-batch, a DQN is able to train the network.
- **Exploration and Exploitation:** In DQN, an agent is able to obtain experience using exploration and exploitation strategies. Using the exploration strategy, the agent is able to explore and learn the environment by taking random actions and achieving the rewards. While in the exploitation strategy, using the current knowledge, DQN maximizes the reward. In DQN, a parameter called epsilon (ε) is the exploration rate that shows the probability of exploring the environment by the agent. During the run time of DQN, this exploration rate will be decayed using a defined decay rate (ε_{decay}) to increase the probability of exploiting strategy.
- **Q-network and Target Q-network:** In DQN, two deep neural networks are created such that Q-network retrieves the Q-function ($Q(s,a)$) as shown in Eq. (8) and target Q-network ($target_Q$) calculates the target value as shown in Eq. (7) using the training process. Target Q-network and its weights (θ') will be updated by Q-network ($\theta' = \theta$) in an interval of time to make the learning process stable. In Eq. (7), $r(s,a,s')$ shows the obtained reward at state s when an action a is selected that changes the state to s'. In addition, $\gamma \in [0,1]$ is the discount factor that demonstrates the importance of the future rewards for the agent such that decreasing the rate will reduce the effect of future rewards on the target value. In this equation, $\alpha \in (0,1]$ shows the learning rate such that the larger learning rate value demonstrates the stronger effect of new data for updating the Q-function as shown in Eq. (8).

$$target_Q \leftarrow r(s,a,s') + \gamma max_{a'} Q(s',a';\theta) \tag{7}$$

$$Q(s,a) \leftarrow (1-\alpha)Q(s,a) + \alpha[target_Q] \tag{8}$$

- **Policy:** In DQN, a policy shows how the agent takes actions to maximize the reward. In the proposed DQN in this paper, the objective is to minimize the network cost in the network presented in [5].

Algorithm 1 shows the process of the proposed DQN algorithm for the proposed linear optimization model in [5]. Table 1 is the descriptions of the inputs of this algorithm. After initializing the variables, Q-network and target Q-network, in each episode in DQN, using the initial observation from the environment and the exploration rate (ε), the agent decides to take an action. Using the value of ε, the agent is able to take a random action from the action space or predict an action using the maximum Q-function value based on the current knowledge. Then, the selected action will be executed in the environment and the next state and the reward will be obtained. As demonstrated in line (13) of Agorithm 1, executing action a also returns a value for *done* which is the condition expression of the *while* loop in line (7). The value of *done* can be returned as *True* in the following situations: 1) if the algorithm reaches a predefined goal reward that can be an expected optimized reward from the environment, or 2) if the *while* loop runs for a predefined number of iterations that can be defined in the experiment based on the type of problem.

Algorithm 1: A DQN algorithm to the linear optimization model of [5]

Input: $G = (V, E), N, N_{ep}, C_{RM}, minC_{RM}, \varepsilon, \varepsilon_{decay}, \varepsilon_{min}, \gamma, \alpha, minbatch_{size}$
Output: N Controller Placements, Controller Assignments, MinCost

1 Initialize the replay memory capacity (C_{RM})
2 Initialize Q-function with random weight θ
3 Initialize target Q-network with weights $\theta' = \theta$
4 **for** *episode 1 to N_{ep}* **do**
5    $done \leftarrow$ False
6   Initialize the first state in the environment s
7   **while** *!done* **do**
8   n $\leftarrow$ generate a random number
9   **if** $n < \varepsilon$ **then**
10   Select a random action a_t from the action space
11   **else**
12        $a = argmax_a Q(s, a; \theta)$
13   Execute action a to observe the next state s' and reward r
       ($r = -[CostFunc(G, s')]$) and return *done*
14   Store transition (s, a, r, s') in the replay memory
15   **if** *bufferSize* $> minC_{RM}$ **then**
16   Sample random minibatch from the replay memory ($minbatch_{size}$)
17   Set $target_Q = \begin{cases} r(s, a, s') & \text{, if } done \\ r(s, a, s') + \gamma max_{a'} Q(s', a'; \theta) & \text{, otherwise} \end{cases}$
18   Update $Q(s, a) \leftarrow (1 - \alpha)Q(s, a) + \alpha[target_Q]$
19   Perform the gradient step on $(target_Q - predicted_Q)^2$
20   reset Q' every C steps by setting $\theta' = \theta$
21      $s \leftarrow s'$
22   **if** $\varepsilon > \varepsilon_{min}$ **then**
23      $\varepsilon = \varepsilon * \varepsilon_{decay}$
24      $\varepsilon = max(\varepsilon_{min}, \varepsilon)$

25 **return** *N Controller Placements, Controller Assignments, MinCost*

Algorithm 2 demonstrates the total network cost for state s' ($Cost_{s'}$) that includes the total cost of setting up flow rules and exchanging configurations (line 2) ($Cost^{s'}_{Setup}$) and the total cost of discovering the network topology by controllers consisting of controller-device and inter-controller communications (line 3) ($Cost^{s'}_{TD}$). The new transition will be stored in the relay memory as shown in line (14). At the next step, if there are enough samples in the replay memory, the training phase starts, including sampling random minbatch from the memory and calculating the target Q-function ($target_Q$) and updating Q-function ($predicted_Q$). Then, the gradient descent is calculated to find the loss.

As mentioned before, the target Q-network is updated in an interval of time. In the next step, the new state will be the current state in the environment. As shown in lines (22) to (24), in each episode, the value of the exploration rate (ε) will be decayed by a decay rate (ε_{decay}) but this value cannot be less than a pre-defined minimum rate (ε_{min}).

Algorithm 2: Cost Function

    Function CostFunc(G, s')
        Calculate $Cost^{s'}_{Setup}$
        Calculate $Cost^{s'}_{TD}$
        $Cost_{s'} = Cost^{s'}_{TD} + Cost^{s'}_{Setup}$
        return $Cost_{s'}$
    End Function

---

4   Experiments, Results and Analysis

To evaluate the performance of the proposed DQN for the proposed CPP in SDMWN, the topology as shown in Fig. 1 is first considered. In this topology, 10 wireless devices are randomly distributed in the environment. The proposed models in the related work [14–16] do not provide details about the values of parameters used for the evaluation. Therefore, in this evaluation, the values are selected from the possible value range for each parameter. The discount factor (γ) and the learning rate (α) values are assumed to be 0.99 and 0.01, respectively, while ε is assumed to be 1 and will be decayed during the iterations by ε_{decays} as 0.99975 and ε_{min} as 0.001. In this paper, AMPL (a mathematical programming language) [20] is used to implement the linear optimization model of [5]. Moreover, we use the NEOS sever [21] to run the model in AMPL using the CPLEX solver [22,23].

Table 2 shows the optimal results for the topology shown in Fig. 1 compared to the results from running the adapted DQN algorithm presented in Algorithm 1 when $N = 3$ and $N_{ep} = 30$. As demonstrated in this table, there is only one difference between the selected controller placements in the optimal solution from the linear optimization model of [5] and the solution of the DQN. The optimal placements are devices 1, 3 and 9, while DQN selects devices 1, 4 and 9 as controllers. Based on the results shown in Table 2, the optimal network cost and the cost achieved from the DQN algorithm are slightly different.

Table 3 presents the selected controllers in the linear optimization model of [5] and the adapted DQN for different numbers of devices. As shown in this table, the selected controllers in both solutions are different in only one controller placement for 10 and 20 devices. Table 4 demonstrates that based on the

Table 1. The definition of symbols used in the adapted DQN presented in Algorithm 1

Name	Description
N	Number of controllers
N_{ep}	Number of episodes
C_{RM}	Size of the replay memory
$minC_{RM}$	Minimum size of the replay memory to start training
ε	Exploration rate
ε_{decay}	Decay rate
ε_{min}	Minimum epsilon value
γ	Discount factor
α	Learning rate
$minbatch_{size}$	minbatch size for training

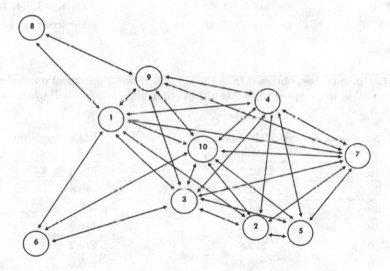

Fig. 1. An SDMWN with 10 wireless network devices

Table 2. Optimal results vs. the results from the adapted DQN algorithm when $N_{ep} =$ 30 and $N=3$

The simplified linear optimization model of [5]		DQN algorithm	
Network cost *(control packets/second)*	Controller placement	Network cost *(control packets/second)*	Controller placement
12.8	Devices 1, 3, 9	14.0	Devices 1, 4, 9

selected controller placements, when increasing the number of devices in the network, compared to the optimal solution, the adapted DQN algorithm is able to find solutions with network cost values close to that obtained from the linear optimization model in [5] for $N = 3$ and $N_{ep} = 30$. It is noted that depending on the network topology, different combinations of 3 placements may have the same network cost.

Table 3. The selected controller placements in the simplified linear optimization model in [5] vs. the adapted DQN Algorithm when $N_{ep} = 30$ and $N = 3$

Number of devices	The linear optimization Model of [5]	DQN algorithm
10	Devices 1, 3, 9	Devices 1, 4, 9
20	Devices 3, 6, 10	Devices 1, 3, 10
30	Devices 1, 8, 15	Devices 1, 22, 30
40	Devices 19, 20, 30	Devices 1, 5, 36
50	Devices 1, 21, 31	Devices 1, 34, 40
60	Devices 1, 17, 19	Devices 1, 2, 10

Table 4. Minimum cost obtained from the simplified linear optimization model in [5] vs. the results of the adapted DQN Algorithm *(control packets/second)* when $N_{ep} = 30$ and $N = 3$

Number of devices	The linear optimization model of [5]	DQN algorithm
10	12.8	14.0
20	45.6	46.8
30	84.4	88.0
40	138.0	148.8
50	213.8	219.2
60	260.4	286.6

Figure 2 and Fig. 3 demonstrate the average number of inter-controller hops and the average number of controller-device hops, respectively when increasing the number of devices in the network obtained from the linear optimization problem of [5] and the adapted DQN algorithm. As shown in Fig. 2 and Fig. 3, the adapted DQN algorithm is able to find the controller placements and assign the controllers to devices such that the average number of hops among devices and their assigned controllers and the average number of hops among controllers in the network are very close to the optimal solution from the linear optimization problem of [5]. As demonstrated in these figures, for a fixed number of controllers

($N = 3$), when increasing the number of devices in the network, the average number of controller-device hops increases that also results in higher network cost as shown in Table 4. On the other hand, these figures illustrate that finding the controller placements in SDMWN could be a trade-off between the average number of inter-controller hops and the average number of controller-device hops in the network. For example, for a network with 60 devices, the adapted DQN algorithm is able to reduce the average number of inter-controller hops (Fig. 2) compared to the optimal solution; however, in this case, the average number of controller-device hops and the network cost as shown in Fig. 3 becomes higher.

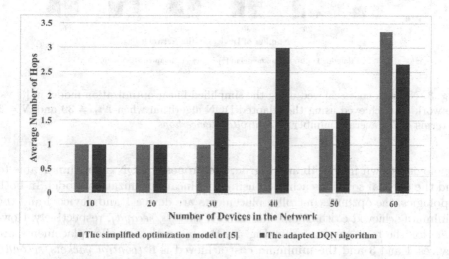

Fig. 2. The simplified linear optimization model in [5] vs. the adapted DQN algorithm when $N_{ep} = 30$ and $N = 3$ in terms of the average number of *inter-controller hops*

To evaluate the performance of the proposed DQN algorithm in SDMWN with topology changes, a scenario is considered as shown in Fig. 4a. The scenario is assumed not to be high mobility. As shown in Fig. 4a, at time t_1, the topology has 6 static network devices. In this topology, the proposed DQN algorithm is able to find device 1 and device 4 as controller placements when $N = 2$ as demonstrated with red circles in this figure. It is assumed that, after a while (Δt), at time t_2 ($t_1 + \Delta t$), device 3 moves in the network and device 3 and device 4 are disconnected as shown in Fig. 4b. At this time, Algorithm 1 selects device 1 and device 4 as controllers in the network. Finally, at time t_3 ($t_2 + \Delta t$), device 5 moves such that link (5, 6) is disconnected and device 5 is in the transmission range of device 1 as shown in Fig. 4c. At this time, running the DQN algorithm finds the same controller placements (devices 1 and 4).

Table 5 shows the controller placements and minimum cost achieved from running the DQN for the aforementioned topologies as shown in Fig. 4. Moreover, Table 6 demonstrates the average number of controller-device hops and the average number of inter-controller hops in these three topologies. In the

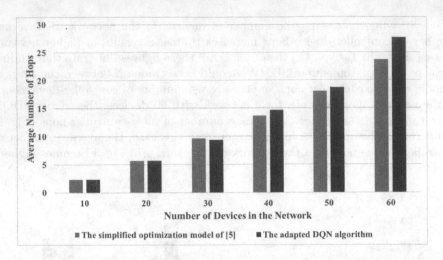

Fig. 3. Network cost achieved using the simplified linear optimization model in [5] vs. Network cost achieved using the adapted DQN algorithm when $N_{ep} = 30$ and $N = 3$ in terms of the average number of *controller-device hops*

topologies shown in Fig. 4b and Fig. 4c, the proposed DQN algorithm is able to find the optimal solutions achieved using the linear optimization model. In both topologies, the optimal controller placements are device 1 and device 4 and the minimum achieved cost is 6 and 5.6 (*control packets/second*), respectively. However, for the topology shown in Fig. 4a, the optimal controller placements are devices 4 and 3 and the minimum cost achieved is 6 (*control packets/second*); the proposed DQN, on the other hand, identifies devices 1 and 4, and the cost is 6.4 for Fig. 4a, as shown in Table IV.

Figure 5 and Fig. 6 present the execution time (*second*) for running the proposed DQN algorithm for topologies shown in Fig. 4 when $N_{ep} = 30$ and increasing the number of iterations used to change the value of *done* in Algorithm 1. As mentioned earlier in Sect. 3, the value of *done* will be *True* if the algorithm finds the goal reward or if *while* loop runs for a pre-defined number of iterations. If the algorithm finds the goal reward before the pre-defined number of iterations that changes the value of *done* to *True*, *while* loop stops. Otherwise, the loop continues for the pre-defined number of iterations. For the topologies shown in Fig. 4, the goal reward is considered as the optimal network cost achieved from the linear optimization model of [5]. As demonstrated in Fig. 5 and Fig. 6, the execution time for the first topology is higher compared to the topologies shown in Fig. 4b and Fig. 4c. In the topologies shown Fig. 4b and Fig. 4c, Algorithm 1 is able to find the goal reward (optimal cost) faster, and hence the iterations

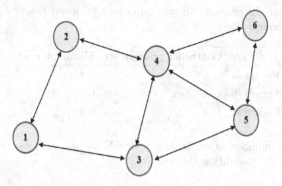

(a) Network topology at time *t1*

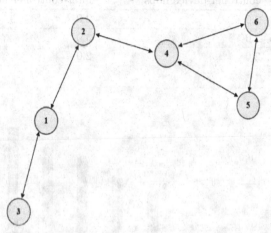

(b) Network topology at time *t2*

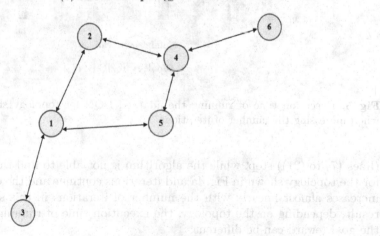

(c) Network topology at time *t3*

Fig. 4. Topology changes in a wireless network with 6 devices

Table 5. Controller placements and minimum cost achieved *(control packets/second)* for the mobility scenario when $N_{ep} = 30$

Topology	Controller placements	Minimum cost
Figure 4a	1, 4	6.4
Figure 4b	1, 4	6
Figure 4c	1, 4	5.6

Table 6. Average number of controller-device hops and average number of inter-controller hops for the mobility scenario

Topology	Avg. number of controller-device hops	Avg. number of inter-controller hops
Figure 4a	2.0	2.0
Figure 4b	2.0	3.0
Figure 4c	2.0	2.0

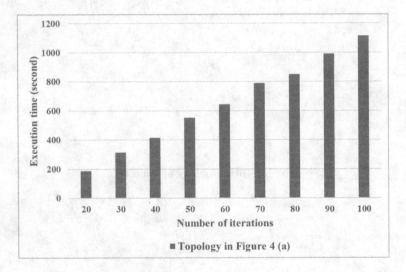

Fig. 5. Execution time of running the adapted DQN for topology shown in Fig. 4a when increasing the number of iterations

(lines (7) to (21)) stop, while the algorithm is not able to find the optimal cost for the topology shown in Fig. 4a and iterations continue and the execution time increases almost linearly with the number of iterations in this situation. As a result, depending on the topology, the execution time of running DQN to find the goal reward can be different.

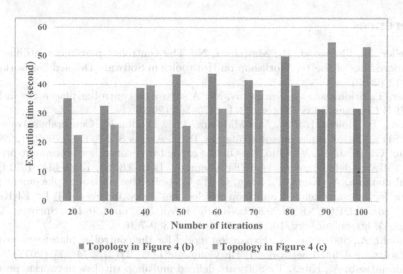

Fig. 6. Execution time of running the adapted DQN for topologies shown in Fig. 4b and Fig. 4c when increasing the number of iterations

5   Conclusion and Future Research

DRL has been applied to various fields, which shows its effectiveness and potential. In this paper, using DRL was investigated to solve the CPP in SDMWN while minimizing the generated control overhead in the control plane referred to as the network cost. To achieve this goal, DQN was adapted to the CPP in SDMWN and we compared the performance of DQN with the optimal solutions. The results demonstrate that DQN is able to find the controller placements and assignments for a given number of controllers while the network costs, the average number of inter-controller hops and controller-device hops obtained from DQN, are close to those of the optical solutions for different networks.

DQN has potential for more complicated problems than algorithm- or model-based approaches, as it is data-based without the need of a complex analytic model. Therefore, adapting DQN in SDMWN with mobile devices when increasing the number of devices and having controller's mobility is being conducted. In addition, investigating the impact of different values for the defined parameters in the adapted DQN on the network performance in SDMWN needs to be addressed in the future research. Finally, there are different factors that have a direct effect on the execution time of the DQN, including the number of episodes, the number of iterations for the while loop and the goal reward. In this case, the execution time of running DQN to solve the CPP in SDMWN should be studied in future.

Acknowledgment. The authors acknowledge the support from the Natural Sciences and Engineering Research Council of Canada (NSERC) through the Discovery Grant program.

References

1. Heller, B., Sherwood, R., McKeown, N.: The controller placement problem. In: Proceedings of the 1st Workshop on Hot topics in Software Defined Networks, pp. 7–12 (2012)
2. Das, T., Sridharan, V., Gurusamy, M.: A survey on controller placement in SDN. IEEE Commun. Surv. Tutor. **22**(1), 472–503 (2020)
3. Isong, B., Molose, R.R.S., Abu-Mahfouz, A.M., Dladlu, N.: Comprehensive review of SDN controller placement strategies. IEEE Access **8**, 170070–170092 (2020)
4. Yao, G., Bi, J., Li, Y., Guo, L.: On the capacitated controller placement problem in software defined networks. IEEE Commun. Lett. **18**,(8), 1339–1342 (2014)
5. Zahmatkesh, A., Kunz, T., Lung, C.-H.: Cost-effective controller placement problem for software defined multihop wireless networks. In: Foschini, L., El Kamili, M. (eds.) ADHOCNETS 2020. LNICSSITE, vol. 345, pp. 130–146. Springer, Cham (2021). https://doi.org/10.1007/978-3-030-67369-7_10
6. Sallahi, A., St-Hilaire, M.: Expansion model for the controller placement problem in software defined networks. IEEE Commun. Lett. **21**(2), 274–277 (2017)
7. Zahmatkesh, A., Kunz, T.: Software defined multihop wireless networks: promises and challenges. J. Commun. Netw. **19**(6), 546–554 (2017)
8. Kumari, A., Sairam, A.S.: Controller placement problem in software-defined networking: a survey. Networks, 1–29 (2021)
9. Roscher, R., Bohn, B., Duarte, M.F., Garcke, J.: Explainable machine learning for scientific insights and discoveries. IEEE Access **8**, 42200–42216 (2020)
10. Arulkumaran, K., Deisenroth, M.P., Brundage, M., Bharath, A.A.: Deep reinforcement learning: a brief survey. IEEE Signal Process. Mag. **34**, 26–38 (2017)
11. Fran cois-Lavet, V., Henderson, P., Islam, R., Bellemare, M.G., Pineau, J.: An Introduction to Deep Reinforcement Learning. Now Publishing Inc., Boston (2018)
12. Goodfellow, I., Bengio, Y., Courville, A.: Deep Learning. MIT Press, Cambridge (2016)
13. Sutton, R.S., Barto, A.G.: Reinforcement Learning: An Introduction. MIT Press, Cambridge (1998)
14. Mostafaei, H., Menth, M., Obaidat, M.S.: A learning automaton-based controller placement algorithm for software-defined networks. In: 2018 IEEE Global Communications Conference (GLOBECOM), pp. 1–6 (2018)
15. Wu, Y., Zhou, S., Wei, Y., Leng, S.: Deep reinforcement learning for controller placement in software defined network. In: IEEE INFOCOM 2020 - IEEE Conference on Computer Communications Workshops (INFOCOM WKSHPS), pp. 1254–1259 (2020)
16. Yuan, T., Neto, W.d.R., Rothenberg, C.E., Obraczka, K., Barakat, C., Turletti, T.: Dynamic controller assignment in software defined internet of vehicles through multi-agent deep reinforcement learning. In: IEEE Transactions on Network and Service Management, vol. 18, no. 1, pp. 585–596 (2021)
17. Mnih, V., Kavukcuoglu, K., Silver, D., et al.: Human-level control through deep reinforcement learning. Nature **518**, 529–533 (2015)
18. Wang, G., Zhao, Y., Huang, J., Duan, Q., Li, J.: A K-means-based network partition algorithm for controller placement in software defined network. In: 2016 IEEE International Conference on Communications (ICC), pp. 1–6 (2016)
19. Lowe, R., Wu, Y., Tamar, A., Harb, J., Abbeel, P., Mordatch, I.: Multi-agentactor-critic for mixed cooperative-competitive environments. In: Proceedings of the 31st International Conference on Neural Information Processing Systems, NIPS 2017, Red Hook, NY, USA, pp. 6382–6393, Curran Associates Inc. (2017)

20. Fourer, R., Gay, D.M., Kernighan, B.: Algorithms and model formulations in mathematical programming. AMPL: A Mathematical Programming Language, pp. 150–151. Springer, Heidelberg (1989). https://doi.org/10.1007/978-3-642-83724-1
21. Gropp, W., More, J.J.: Optimization environments and the NEOS server. In: Buhmann, M.D., Iserles, A. (eds.) Approximation Theory and Optimization, pp. 167–182. Cambridge University Press, Cambridge (1997)
22. CPLEX: IBM's Linear Programming Solver. http://www.ilog.com/product/cplex/. Accessed May 2021
23. NEOS Server for CPLEX/AMPL. https://neos-server.org/neos/solvers/lp:CPLEX/AMPL.html. Accessed May 2021

Distance Estimation Using LoRa and Neural Networks

Mira Abboud[1,3], Charbel Nicolas[2,3], and Gilbert Habib[3(✉)]

[1] Department of Computer Science, University of Prince Edward Island, Cairo, Egypt
mira.abboud@uofcanada.edu.eg
[2] Computer Science Department, CNAM, Paris, France
{charbel.nicolas,gilbert.habib}@ul.edu.lb
[3] LaRRIS, Faculty of Sciences, Lebanese University, Fanar, Lebanon

Abstract. Disasters like floods, avalanches, earthquakes are one of the main causes of death in human history. Search and rescue operations use drones and wireless communication techniques to scan and find the location of victims under rubble. Renowned for its resilience to the different causes of signal attenuation, LoRa wireless communication has been considered as the best candidate to employ for this type of operations. Thus, in this paper, we present a solution based on LoRa radio parameters and Artificial Neural Networks to estimate the distance between the rescue drone and the victim. By using real measurements that represent an actual search and rescue operation, we have achieved distance estimations (between 0 to 120 m) with less than 5% mean error. Add to this, our results, which are based on various LoRa radio parameters, show an improvement of 78% over the mechanisms that use RSSI as the only parameter.

Keywords: Internet of Things · Search and rescue · LoRa · Hybrid indoor and outdoor communication · Indoor and outdoor environments · Neural networks · Artificial intelligence

1 Introduction

Accurately localizing victims under rubble in natural disasters (e.g. floods, avalanches, earthquakes, etc.) is not an easy task even when using dedicated equipment (e.g. ARVA and RECCO cf. Fig. 1) [7].

Other equipment uses wireless technologies like GSM, 5G, WiFi, and Bluetooth to determine the location of the victim. However, obstacles in a disaster can cause signal blocking and low level of penetration in these types of wireless technologies; which limits thereby the device's precision in distance estimation. In a recent study, Judy et al. [4] shows that LoRa technology is one of the best wireless technologies to be used in concrete buildings. Due to its modulation technique, frequency band and Forward Error Correction (FEC) LoRa provides resilience to the different causes of signal attenuation. Moreover, LoRa module is becoming ubiquitous and a dominant technology used in the IoT networks and it is considered a promising technology developed to overcome the signal attenuation problem.

© Springer Nature Switzerland AG 2022
É. Renault et al. (Eds.): MLN 2021, LNCS 13175, pp. 148–159, 2022.
https://doi.org/10.1007/978-3-030-98978-1_10

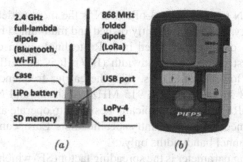

(a)   *(b)*

Fig. 1. cf. [7] shows a: Lopy-4 board and b: Reference ARVA: PIEPS POWDER BT

Determining the distance between two wireless nodes could be achieved using path loss models. This latter depends on the topology and existing obstacles and could be a medium to extract the distance. For instance, using the following path loss formula:

$$Transmission\ power = Received\ power + Path\ loss\ power \qquad (1)$$

$$Pathloss\ power = 20\log_{10}\left(\frac{4\pi d}{\lambda}\right) + c \qquad (2)$$

λ *is the wavelength, d is the transmitter−receiver distance in the same units as the wavelength and c is a cons*tan*t that depends on the system configuration.*

the path loss power could be readily calculated (2); meaning thereby that the distance could be theoretically estimated from (2):

$$d = \frac{\lambda 10^{\left(\frac{PL-c}{20}\right)}}{4\pi} \qquad (3)$$

PL Is the path loss power

However, the mathematical formulas do not give precise values since the path loss model depends on many variables that are mainly related to the transmitter and receiver configurations, and to the communication environment. Therefore, we are proposing in this paper an adaptable path loss model that is found by modeling an artificial neural network trained and tested on a real dataset.

The rest of this paper is organized as follows. Section 2 presents thoroughly the characteristics of LoRa radio. In Sect. 3, we discuss the related work. In Sect. 4, we introduce a detailed description of the artificial neural network. We show in Sect. 5 the experimental setup. The results obtained from the experiment are discussed and analyzed in Sect. 6. We reach our conclusions in Sect. 7 and present our future work.

2 LoRa Technology

Low Power Wide Area Network (LPWAN) is intended for battery-operated devices to send small data packets wirelessly over long distances. LPWAN technologies prove useful in monitoring utilities, home automation, fleet management, and agriculture. Among the numerous LPWAN technologies, LoRaWAN is the most widely adopted. It builds upon LoRa technology, which is currently owned and maintained by Semtech.

Many parameters can be configured in LoRa technology in order to improve its performance. The first one is the bandwidth (BW). It is the difference between the channel's upper and lower frequencies, with the carrier frequency centered between them (433 MHz, 868 MHz "Europe", 915 MHz "Australia and North America" and 923 MHz "Asia"). LoRa devices are typically designed to operate at bandwidths of 125, 250, and 500 KHz to meet different country specifications. European specifications allow 125 and 250 KHz channel bandwidths only.

Another interesting parameter is the spreading factor (SF) which represents the number of bits that can be represented by each symbol. 2^{SF} values can be represented on each symbol, and typically the SF value spans between 7 (128 chips/symbol) and 12 (4096 chips/symbol). A higher SF results in more energy consumption and a higher time on air [4]. However, a higher SF increases the signal to noise ratio (SNR) limit and should therefore allow for a longer distance transmission.

The duration to transmit a symbol is affected directly by the bandwidth and spreading factor. The symbol duration is calculated using the formula below:

$$T_{Symbol} = \frac{2^{SF}}{BW} \tag{4}$$

Notice that doubling the bandwidth halves the symbol duration, whereas increasing the spreading factor by one doubles the symbol duration.

FEC is integrated into LoRa's proprietary modulation technique. FEC allows error detection and to some extent error correction. Since in LoRa retransmissions are not encouraged, FEC is desirable. The ratio of useful bits to total bits is known as the coding rate, it can take values of 4/5, 4/6, 4/7, and 4/8. In what follows, we will use CR to refer to the coding rate's denominator. The coding rate is of interest to us, and changing it will modify the packet's airtime.

RSSI stands for Received Signal Strength Indicator. It represents the power of a received signal on the receiver side. RSSI characterizes the attenuation of radio signals during propagation. Therefore, it has a strong relation with the distance traveled by the signal. RSSI can have an important role in estimating the distance between the sender and the receiver.

SNR stands for signal-to-noise ratio, it is the ratio between the power of a signal and the power of the background noise. This parameter, measured by the receiver, will also help to estimate the distance.

3   Related Work

In literature, different studies to determine distances or locations using information provided by LoRa devices have been presented. Moreover, many focused on the "Search and Rescue" application and developed algorithms for distance estimation using the RSSI to model the distance e.g. [6, 7]. In this type of application, the difficulty persists in the fact that the sender or receiver is under an obstacle that degrade drastically the signal power. Most research work used Antwerp database e.g. [2, 3, 5] to estimate distances, which can be used for outdoor localization only not for the case of a victim under rubble or avalanche (case of Search and Rescue). Therefore, any pathloss model that does not take into consideration the hybrid topology, where a node is covered (indoor) by different types of obstacles and the other in the open (outdoor), will give imprecise estimation of the pathloss and by that, imprecise distances. To our knowledge, none used the neural network and LoRa radio parameters as input to improve the distance estimation in case the emitter or receiver exists in heterogeneous environments (indoor/outdoor).

The energy consumption constraints along with the inability to use GPS indoors pushed further the studies on using the RSSI for estimating distances. In [8] several techniques like trilateration, multilateration, triangulation, and fingerprinting using the RSSI have been presented. Nevertheless, the RSSI value is influenced by any modification made for any of these transmission parameters: the spreading factor (SF), transmission power and the time on air (TOA) as shown in [1, 4] and thus using RSSI as the only parameter for the estimation is considered a limitation in the distance estimation area. Although, in [3, 5] only the RSSI was used as input for their proposed algorithms and presented promising results for determining the distance using Artificial neural network (ANN), they failed to consider the effect of varying the configuration and the parameters of the radio module on the RSSI value. We argue that using only the RSSI as input to the distance predictability model used will drastically affect the RSSI measurements and by that will degrade the distance precision.

The authors in [6, 7] worked on the search and rescue operation of LoRa radio. In [7], the RSSI was used with a static pathloss model, and they proposed a fixed Spreading factor SF value equal to 12 to predict the distance. This configuration is not dynamic or efficient because, based on the work done in [9] the SF should be dynamic to optimize the packet delivery ratio. Moreover, based on our work in [4], using the SF = 12 for a long duration is not recommended due to the higher Time on Air, hence, a higher collision probability. Moreover, the static pathloss model is not applicable in different scenarios and topologies. In [6], the authors proposed an improvement on the use of land operated LoRa by using a drone-aided system to improve localization in LoRa IoT networks. Their setup experiment is shown in the Fig. 2.

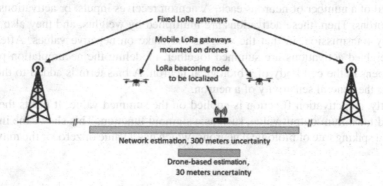

Fig. 2. The experimental setup used in [6]

The drone-aided system proved efficient in specific topologies due to the static path loss model used. In a dynamic and heterogeneous environment, a dynamic path loss model is needed to adapt, this is why we propose to implement a dynamic model that adapts to LoRa's configuration and the diverse states of the topology.

4 Neural Network

Through their power and scalability, neural networks have become the defining model of deep learning. Neural networks are composed of neurons, where each individual neuron takes an input, performs only a simple computation and provides an output. The power of a neural network comes instead from its imitation of the human brain behavior and architecture; more specifically, it comes from the complexity of the connections that a large number of neurons can form (Fig. 3).

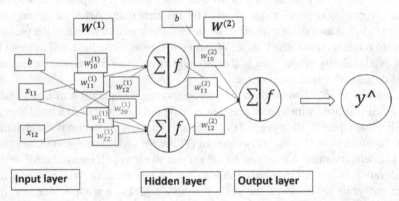

Fig. 3. A Neural Network composed of 1 hidden layer with two neurons

First, a deep neural network is composed of one or more hidden layers, that in turn are composed of a number of neurons each. A neuron receives inputs, or activations, from other neurons. Then, these activations are multiplied by weights, and they also model inhibitory transmission, in that the weights may take on negative values. Afterward, these weighted activations are summed together, modeling the accumulation process that happens in the cell body of a biological neuron. A bias term is added to the sum, modeling the general sensitivity of a neuron.

Finally, an activation function is applied on the summed value. It limits the minimum and maximum output value, such as a sigmoid function. This shows the intrinsic minimum spiking rate of biological neurons, which has a value of zero, or the maximum rate.

The Loss Function

Machines learn using the loss function. It is a method of evaluating how well specific algorithm models that given data. If predictions differ too much from original results, the loss function would generate a very large number. In a neural network, the Loss function is applied to calculate the gradients, and gradients are used to update the weights of the Neural Net.

The Activation Function

An activation function is used to get the output of a node. We use it with neural networks to determine their output. It maps the resulting values in a range between the two integer values 1 and −1.

The Optimizer

Optimizers are the algorithms used to update the weights of the neural network in a manner that reduces the losses; they are responsible for providing the most accurate solutions. Optimizers are said to be efficient if they help to overcome the problem of local optimum, overshooting and vanishing. One of the best optimizers is Adam mini-batch gradient descent while using dynamic learning rates.

5 Experimental Setup

In our experiment, we aim to build a neural network model in order to estimate the distance between the sender and the receiver. This model uses as input the SF, CR, RSSI, SNR, and the estimation distance as output. We assume that the transmission power is fixed to a value that does not deplete rapidly the search and rescue drone from power.

For this task, we chose the TTGO LoRa32-OLED board with its built-in linear antenna with a carrier frequency of 868 MHz. In addition to LoRa, these boards come with Bluetooth and WiFi, providing flexibility for expanding our experiments in the future. Moreover, these boards can be programmed within the ubiquitous Arduino IDE, which facilitates the task of configuring them and logging their collected information. The boards' OLED screen makes them easy to monitor on the go. We fixed the transmit power at 10 dBm and the bandwidth to 125 kHz. The reception sensitivity depends on SF according to the TTGO datasheet and LoRa specifications.

A fixed receiver (represents the victim) placed inside a building and multiple transmission points were selected for the sender (represents the rescue drone) as shown in the figure below; The distance between the two devices is known, it varies between 0, 10, 20, 30, 40, 50, 60, and 120 m (Figs. 4 and 5).

Fig. 4. The building structure (Math building) at the Lebanese University faculty of sciences branch II

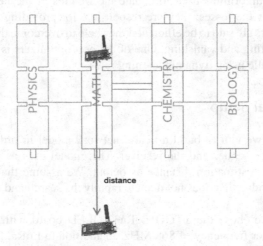

Fig. 5. The topology of the experimental setup

The building walls are made of reinforced concrete. The wall's thickness varies between 15 and 25 cm. The doors and windows are made of steel and glass. The indoor LoRa device is located in the middle of the corridor.

At every test point, we test all the combinations of spreading factors SF = {7,8,9,10,11,12}, coding rates CR = {5,6,7,8}. For every combination, the sender sends a Beacon packet, the receiver measures the RSSI hereafter called B_RSSI and SNR hereafter called B_SNR of the received packets. The receiver encapsulates in a packet these two values and replies back to the sender. This latter will also measure the RSSI and SNR called R_RSSI and R_SNR respectively. The sender will save all the measured data and the Lora parameters (SF, CR) in a table.

6 Results and Analysis

The Dataset

The dataset contains 32,843 observations (data row) collected from the experiment described in Sect. 5. Table 1 describes the distribution of each parameter of the collected data, whereas Table 2 provides the average of each parameter according to the different distances.

Table 1. The distribution of the different inputs

	SF	CR	b_rssi	b_snr	r_rssi	r_snr
Mean	9.53	6.49	−94.29	7.51	−86.19	6.77
Std	1.70	1.12	20.92	5.23	20.39	6.63
Min	7.00	5.00	−141.00	−15.00	−118.00	−19.75
25%	8.00	5.00	−110.00	7.00	−101.00	4.75
50%	10.00	6.00	−103.00	9.00	−95.00	9.25
75%	11.00	7.00	−78.00	11.00	−71.00	11.50

Table 2. The average of each parameter according to the different distances

Distance	SF	CR	b_rssi	b_snr	r_rssi	r_snr
0	9.50	6.50	−55.87	10.67	−48.02	11.60
20	9.51	6.50	−74.90	10.80	−67.81	11.15
30	9.50	6.50	−92.15	9.81	−84.63	10.82
40	9.50	6.50	−104.66	9.22	−96.41	9.60
50	9.50	6.50	−107.95	7.47	−99.29	4.66
60	9.51	6.50	−109.41	7.30	−100.74	4.87
120	9.71	6.41	−118.15	−4.19	−109.38	−7.10

The Neural Network Architecture

Regarding the data preparation, we applied the Pareto Principle (20/80) to split our dataset. The bucket of samples is taken almost equal for all the distances in each split. The training dataset was cleaned from the null value by removing outliers. These latter were removed under the condition that the RSSI is less than −150. A limited number of observations (~10 were removed), while the testing dataset was kept as it is.

In order to avoid vanishing and exploding problems, we initialize the parameters using the normal kernel initializer; accordingly, normal distribution was used to initialize the weights.

As highlighted in Tables 3 and 4, several experiments with different numbers of layers and neurons have been performed in order to find the most accurate results. More specifically, the first set of experiments is composed of 6 neurons input layer; which implies that the 6 parameters were the NN inputs. Table 3 summarizes the experiments and the results. Whereas in the second set of experiments, depicted in Table 4, only b_rssi and r_rssi were taken as inputs to train the model.

Table 3. Neural Network parameters and results for 6 neurons input layer (SF, CR, R_RSSI, B_RSSI, R_SNR, B_SNR as inputs)

Model - results	Training loss (MSE)	Validation loss (MSE)	Mean error
2 Layers, 64 neurons	20.74	22.72	2.72
3 Layers, 64 neurons	14.45	13.36	1.74
3 Layers, 128 neurons	10.14	10.47	3.23
4 Layers, 64 neurons	9.46	10.25	1.14
4 Layers, 128 neurons	9.27	8.49	0.81

Table 4. Neural Network parameters and results for 2 neurons input layer (R_RSSI and B_RSSI as inputs)

Model - results	Training loss (MSE)	Validation loss (MSE)	Mean error
2 Layers, 128 neurons	68.41	65.96	4.32
3 Layers, 64 neurons	73.85	73.16	4.48
3 Layers, 128 neurons	57.78	55.23	3.45
4 Layers, 64 neurons	60.49	57.96	3.63
4 Layers, 128 neurons	59.88	57.14	3.67

Model Training

Since MSE (Mean Squared Error) ensures that a trained model has no outlier predictions with huge errors due to the squaring part of the function, we chose it as a loss function to train the model. Add to this, we applied the activation function reLu on each Neuron, which has a linear behavior and is mostly used to predict a continuous positive value. For the optimizers, Adam gave us the best results, with a batch size = 32. The batch size 32 allowed us to run the simulation faster than taking one observation at each iteration; add to this, it helped to obtain better results which could be due to the minimization of the effect of the outliers observations on the parameters optimization. After 200 epochs, which means 200 passes on the dataset, we obtained a validation loss of 8.49. The graph of Fig. 6 shows the decrease of losses (validation and training) and how it ends with saturation.

Note that the validation loss is the loss (according to the MSE function) between the real value and the predicted value of the testing data; whereas the train loss is applied on the training data.

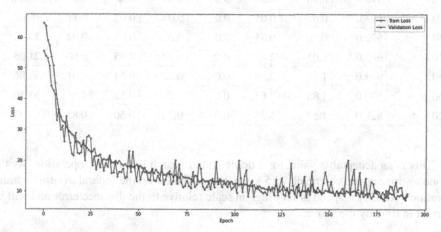

Fig. 6. The training loss and validation loss values during the training phase on 200 epochs

Results Analysis

First, the results prove that by using the Neural Network model we can predict the distance of an object with high accuracy.

Moreover, our hypotheses was validated by the results presented in Tables 3 and 4. The results show that using the six different parameters and metrics from LoRa messages as input showed 78% improvement on distance accuracy over only using the RSSI metric.

Table 5 shows the mean error on the different distances. For example, for the distance 60 m, 50% of the data has an error less than 1.27 m, 75% of the data has an error of less than 1.56 m; for the distance 120 m, 75% of the data has an error less than 0.85. By slicing the column of 75%, we can conclude that 75% of the predictions has an error less than one percent, which is negligible regarding the large distances taken.

Even though the max column shows a high error, the number of observations that achieved the max value is very limited (one or two instances per distance). This is approved by the low mean and standard deviation values. We believe that not applying the cleaning steps, on the testing data, might be the reason for the high maximum.

Figure 7 shows the relationship between the error value and the distance between the LoRa devices. The higher the distance the higher the standard deviation. This is caused by the low RSSI level relatively to the floor noise.

Moreover, the University building is located in a heavily populated and industrial area where various sources of interference have been detected prior to the experiment. Those sources had none negligible signal power relatively to our equipment.

By comparing the standard deviation of the error and the distance, we conclude that the standard deviation is always less than 7% of the distance.

158   M. Abboud et al.

Table 5. Error results summary

Distance	Count	Mean	Std	Min	25%	50%	75%	Max
0	979.0	0.00	0.00	0.0	0.00	0.00	0.00	0.00
20	963.0	0.07	0.05	0.0	0.03	0.05	0.11	0.69
30	922.0	0.10	0.85	0.0	0.02	0.02	0.02	13.88
40	983.0	1.01	2.43	0.0	0.17	0.35	0.66	21.05
50	963.0	1.27	2.60	0.0	0.23	0.54	1.01	50.86
60	930.0	1.83	3.65	0.0	0.20	0.52	1.56	53.50
120	829.0	1.53	5.25	0.0	0.26	0.50	0.83	63.37

This is an acceptable value for a drone added search and rescue operation. As the drone average speed can reach 72.5 km per hour (based on the federal aviation administration FAA). This speed is very high in scale relative to the distance error and will be correct in time by the drone.

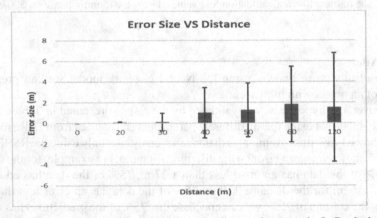

Fig. 7. Shows the error value relatively to the distance between the LoRa devices

7   Conclusion

In this paper, by using our own testbed measurements, we present a method that estimates the distance between two LoRa devices for the search and rescue operation. This method is based on the artificial Neural Network.

The results show that the improvement reached 78% over the mechanisms that rely only on the RSSI as the input parameter. Moreover, the mean error does not exceed 5%, the standard deviation is always less than 10% of the distance and the Mean error of 50% of the trial is less than 1%, this shows the precision efficiency the neural network model used to predict the distance between the victim and the rescue team.

Focusing on the following two axes is our near future work: (1) Estimate the distance while having the objects in motion. (2) Applying more complex models while trying to not fall into the problem of overfitting. More specifically, we will enlarge the architecture with some dropout percentages and regularization techniques.

Acknowledgements. This work was supported by the Lebanese University under Grant 2019 and partially funded by the ANR INTELLIGENTSIA (nb. ANR-20-CE25-0011).

References

1. Ayele, E., Hakkenberg, C., Meijers, J.P., Zhang, K., Meratnia, N., Havinga, P.M.: Performance analysis of LoRa radio for an indoor IoT applications. In: International Conference on Internet of Things for the Global Community (IoTGC) (2017)
2. Daramouskas, I., Kapoulas, V., Paraskevas, M.: Using neural networks for RSSI location estimation in LoRa networks. In: IISA (2019)
3. Ingabire, W., Larijani, H., Gibson, R., Qureshi, A.: Outdoor node localization using random neural networks for large-scale urban IoT LoRa networks. In: MDPI Algorithms, vol. 14, no. 11 (2021)
4. AbiNehme, J., Nicolas, C., Habib, G., Haddad, N., Duran-Faundez, C.: Experimental study of LoRa performance: a concrete building case. In: 2021 IEEE International Conference on Automation/XXIV Congress of the Chilean Association of Automatic Control (ICA-ACCA) (2021)
5. Nguyen, T.: LoRa Localisation in Cities with Neural Networks (2019)
6. Delafontaine, V., Schiano, F., Cocco, G., Rusu, A., Floreano, D.: Drone-aided localization in LoRa IoT networks. In: IEEE International Conference on Robotics and Automation (ICRA) (2020)
7. Bianco, G.Ma., Giuliano, R., Marrocco, G., Mazzenga, F., Mejia-Aguilar, A.: LoRa system for search and rescue: path loss models and procedures in mountain scenarios. In: IEEE Internet of Things Journal (2021)
8. Laoudias, C., Moreira, A., Kim, S., Lee, S., Wirola, L., Fischione, C.: A survey of enabling technologies for network localization, tracking, and navigation. In: IEEE Communications Surveys & Tutorials (2018)
9. Haiahem, R., Minet, P., Boumerdassi, S., Azouz Saidane, L.: Collision-free transmissions in an IoT monitoring application based on LoRaWAN. In: Sensors, MDPI, 2020, ff10.3390/s20144053ff. ffhal-02908985

Author Index

Printed in the United States
by Baker & Taylor Publisher Services